CONTENTS

LIST OF CONTRIBUTORS

Sheldon Baron

Bolt Beranek and Newman Inc.
Cambridge, Massachusetts

Elliot E. Entin

ALPHATECH, Inc.
Burlington, Massachusetts

Ruston M. Hunt

Search Technology, Inc.
Atlanta, Georgia

David L. Kleinman

Department of Electrical
Engineering and Computer Science
University of Connecticut and
ALPHATECH, Inc.
Burlington, Massachusetts

Adolfo Lagomasino

Department of Systems Engineering
University of Virginia

Krishna Pattipatti

ALPHATECH, Inc.
Burlington, Massachusetts

Jens Rasmussen

Riso National Laboratory
Roskilde, Denmark

William B. Rouse

Center for Man-Machine
Systems Research
Georgia Institute of Technology and,
Search Technology, Inc.
Atlanta, Georgia

Andrew P. Sage

Department of Systems Engineering
University of Virginia

Thomas B. Sheridan

Department of Mechanical
Engineering
Massachusetts Institute of
Technology

Joseph G. Wohl

ALPHATECH, Inc.
Burlington, Massachusetts

INTRODUCTION

Man-machine systems is an interdisciplinary field of study that draws significantly on concepts and methods of psychology, systems engineering, computer science, management, and related disciplines. Because of this breadth, advances in the field emanate from a wide variety of sources and are published in a rather heterogeneous range of journals, monograph series, etc. As a result, it can be quite difficult to assess the state of art in man-machine systems. The goal of this annual series, *Advances in Man-Machine Systems Research*, is to contribute to ameliorating this difficulty.

Each volume in the series will contain several substantive chapters which report new theoretical or empirical results, or novel applications, on topics such as human decision making and problem solving, human-computer interaction, human monitoring and control of dynamic processes, and basic human performance issues associated with these topics. Each chapter will be a report of original research, typically involving the synthesis of a series of smaller efforts into an overall perspective on a topic. Chapters which primarily provide reviews of the literature will be the exception rather than the rule, as will be reports of any single theoretical or empirical study which could be published as a single journal article. Chapters which explicitly consider the applicability of the results reported to analysis, design, and/or evaluation of man-machine systems will be particularly encouraged.

While there is no fixed plan to devote each volume to a special theme, this inaugural volume deals primarily with human decision making and problem solving in the operation and maintenance of complex dynamic systems. The application areas discussed include electronic troubleshooting, aircraft maintenance, remote manipulation, power plant operations, and military command and control. The methodological perspectives range from systems engineering to artificial intelligence to psychology. However, despite this variety, there are several common themes that repeatedly emerge. These include the human's role as a supervisor of automatic systems, the human's abilities to deal with failure situations, and the overall impact on man-machine systems of current trends in information technology and software engineering. At least at the conceptual level, there appears to be a clear consensus among the authors represented in this volume that these issues will become increasingly important.

William B. Rouse
Editor

A CONTROL THEORETIC APPROACH TO MODELLING HUMAN SUPERVISORY CONTROL OF DYNAMIC SYSTEMS

Sheldon Baron

ABSTRACT

The application of modern control theoretic methods to modelling human supervisory control performance is described. First, specific models developed from this perspective are reviewed to provide the context for a generic modelling approach; these include models for continuous control, monitoring and decision-making and multi-task supervisory control. Then a generic, conceptual supervisory control model based on a control-theoretic viewpoint is discussed in some detail. This model shares many features with the models reviewed previously, but also includes fundamental extensions and generalizations. Finally, the approach is summarized and discussed in terms of some of its advantages and limitations and the needs for additional research.

Advances in Man-Machine Systems Research, Volume 1, pages 1–47.

I. INTRODUCTION

Human supervisory control has been a subject of increasing interest and importance in recent years as the role of operators of complex systems has evolved as a result of the widespread applications of computer technology. Thus, two NATO conferences [Sheridan and Johannsen, 1976; Rasmussen and Rouse, 1980] were devoted to various aspects of the subject and a recent workshop sponsored by the Nuclear Regulatory Commission (1982) was aimed largely at modelling human cognitive activities in connection with the supervisory control tasks associated with operation of nuclear power plants. Though it is extremely difficult to give a precise, mathematical formulation of supervisory control that covers all the system and operator behaviors potentially involved,[1] it is possible to provide a qualitative description of the problem that is useful for our purposes and is based on the various issues discussed in the literature. (The most commonly referred to description of supervisory control is that of Sheridan [1976].)

Although specific applications will emphasize different portions of the supervisory control problem, it is widely agreed that from the operator's viewpoint these problems are generally characterized by the following features:

- large scale, technological systems involving high economic value and, often, significant risk
- complex and dynamic processes with many outputs to be "controlled" and many potential inputs for achieving that control
- a structure in which there are many sub-systems, with the coupling between variables in different sub-systems much looser than that among variables in a given sub-system
- a significant degree of automation, both in system monitoring and control
- relatively slow response of the variables to be controlled (with rapidly changing variables controlled automatically)
- event driven demands
- the need to interact and coordinate with other operators and/or external entities, and
- the requirement to follow specific procedures in defined situations (available in procedures "manuals" or residing in the operator's memory).

Examples of systems/problems that fall within the domain of supervisory control are: the control of large industrial processes (e.g., chemical or power plants); flight management problems; air traffic control; supertanker control; and the management and control of certain weapons systems.

Thus, the operator engaged in supervisory control will generally be immersed in a multi-task, multi-objective, multi-person environment which emphasizes cognitive behavior rather than psychomotor performance. For these problems, the operator's functions or activities include planning, monitoring, situation assessment, decision-making (selection of appropriate tasks and procedures), control (usually discrete but sometimes continuous) and communication. While most applications will involve all these activities, the relative importance of each will depend on the specific problem.

The problem for designers of such complex *man-machine* systems, and, often, for those charged with setting standards for their operation, is how to analyze them to ensure that they function properly (i.e., that the controls, displays, procedures and training, are adequate to guarantee safe operation in the face of inevitable disturbances and breakdowns while maintaining efficient operational performance in the normal mode). Frequently, the problem reduces to one of adequately accounting for the human element in such a system.

The application of modern man-machine systems analysis techniques will undoubtedly be required to account fully for the "human-in-the-loop" in such situations. There is a continuum of tools that can aid this systems analysis process. These range from task analysis to full-scale simulation and test of alternative designs and procedures. Task analysis fails to provide the desirable quantitative basis for regulation and it is virtually impossible to test out all realistic design alternatives in either simulators or real systems. One analytic tool that falls between these methods and that can retain many of their advantages, while avoiding their difficulties, is modelling of the human-machine system.

The importance of "person-in-the-loop" modeling has long been recognized, and many attempts have been made to estimate the operator's effect on system performance and reliability. However, the nature of the desired model has been subject to serious debate, both at the general level of qualitative vs. quantitative models and at the specific level of model definition. As Singleton (1976) has noted, qualitative models can be quite useful but they invite ambiguity and are limited in the range of predictions and evaluations they support. Quantitative models, on the other hand, must be explicit and can predict subtle and important human/system interactions, but they can be difficult and costly to develop and are generally complex and therefore hard to understand. Establishing the validity of either type of model is extremely difficult at best. Inasmuch as quantitative models provide "data" they can be subjected to more detailed validation and, indeed, they should be.

It is not the intent of this chapter to review the various alternative approaches to modelling human supervisory control (for such reviews, see Pew, Baron, Feehrer and Miller, 1977, or Pew and Baron, 1982). Instead, the aim is to describe a control-theoretic approach to the problem that has been emerging in recent years. In this context, it is important to realize that control-theoretic modelling of human performance has matured beyond the relatively simple "human-as-servomechanism modelling" that characterized its early years to the point where tasks involving multivariable control, monitoring and decision-making have been modelled successfully (e.g., Baron, 1976; Kleinman and Curry, 1977; Levison and Tanner, 1971; Gai and Curry, 1976; Pattipati and Kleinman, 1979; Wewerinke, 1981). Recently, efforts have focused on models for supervisory control in multi-task environments that involve a broad range of operator behaviors and, in some cases, interactions with other operators (Muralidharan and Baron, 1980; Baron, Muralidharan, Lancraft and Zacharias, 1980; Zacharias, Baron, and Muralidharan 1982). Several of these models are reviewed in Section II of this chapter as a prelude to the discussion, in Section III, of a generic, conceptual model for supervisory control. The final section of the chapter summarizes the approach and its features and presents some concluding remarks.

II.   SOME CONTROL-THEORETIC MODELS

To provide a background for the approach to modelling supervisory control that we are investigating, it is useful to examine some earlier models that have provided the perspective and impetus for the approach. Of particular relevance are the Optimal Control Model (OCM) of the human operator and some related models for various types of human decision-making.

A.   The Optimal Control Model for Continuous Control

The OCM has been well documented (see Baron and Levison, 1980 for a recent review and an extensive bibliography) so we will only focus on the factors that are important for supervisory control. A block diagram for the OCM is given in Figure 1. This diagram is somewhat different than that usually shown in the literature to emphasize its connection to the supervisory control modelling approach to be discussed later. In particular, the model for the human operator has been separated into three parts: a display processor, an information processor and a procedure processor.

The "display processor" block is the monitoring portion of the model. It contains the perceptual processing limitations of observation noise and

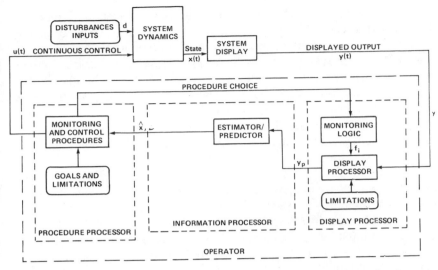

Figure 1. The Optimal Control Model (An alternate view).

time-delay that are included in the OCM. Note that the observation noises are affected by the monitoring logic as defined by the "procedure processor" block. This logic reflects attention-sharing limitations incorporated in the OCM via a model for task-interference (Levison, Elkind and Ward, 1971).

The "information processing" block contains the Kalman estimator/predictor the function of which is to estimate the current state of the system on the basis of the perceived outputs. This limited form of "situation assessment" is all that is required to perform the control task in an "optimal" fashion under the assumptions normally given in deriving the OCM. Moreover, it models the (cognitive) process whereby the operator constructs a set of expectancies concerning the state of the system on the basis of his understanding of the system and his incomplete knowledge of the moment-by-moment state as accessible to him from limited and noisy observations.

The "procedure processor" block incorporates both the monitoring and control strategies or procedures. These are chosen so as to accomplish the desired performance objective, which in the OCM is that of minimizing a cost that is a weighted sum of output and control variances. The "control procedure" is simply a linear feedback law that generates a continuous commanded control; i.e., the commanded control is a linear combination of the estimated state variables where the coefficients are chosen to minimize the cost. These commanded controls are modified by the human's motor response limitations before they are input to the system. The monitoring decision involves choosing the "fractions of attention" that are to be devoted to each display over the control interval so as to optimize the same basic criterion used in computing the control strategy.

The OCM has proven to be capable of predicting, or matching, human performance with remarkable fidelity in a variety of continuous control tasks. It is the author's view that this success derives from a number of basic modelling principles and some specific model constructs that are directly relevant to the development of models for supervisory control.

On the basic level, we note that the OCM is a closed-loop, man-machine system model. It embodies the idea that, to model human performance, one must model the system in which that performance is embedded. Human behavior, either cognitive or psychomotor, is too diverse to model unless it is sufficiently constrained by the situation or environment; however, when these environmental constraints exist, to model behavior adequately, one must include a model for that environment. In addition, the OCM reflects several important trends from modern control theory, in particular the following:

- a state-space description of the system with a corresponding emphasis on internal, rather than input-output, models of system behavior
- direct incorporation of stochastic aspects of the problem (including those of the human)
- optimization of a performance criterion
- consideration of state estimation as integral, and often essential, to the control process
- time-domain analysis

Formulation of the continuous control problem within the framework provided by these concepts has led to a natural way of incorporating human limitations at the processing level (e.g., observation noise and time delay), rather than at the performance level as is done in many human performance models (such as network models [Siegel and Wolf, 1969] or reliability models [Swain and Guttman, 1980]). Moreover, the formulation of the control problem as an optimization problem leads directly to the requirement to include in the controller explicit knowledge of the system dynamics and disturbance statistics, as well as implicit knowledge of the control criterion (to compute the feedback gains); i.e., to the inclusion of appropriate internal models.[2] Because of these factors, the OCM has demonstrated a significant capability for predicting both the goal-oriented control activity and the random response of operators; and, it has done this across a variety of changes in task parameters, such as changes in system dynamics, displays and input disturbances with a relatively invariant set of parameters and without resorting to a subsidiary set of verbal adjustment rules.

In terms of specific model elements, the most significant portion of the OCM for purposes of modelling supervisory control is the estimator which represents the information processing activity of the human in this task.[3] In the OCM, the estimation process is separated from the control behavior in a manner that allows it to be considered independently. This fortuitous separability, which parallels the separation between perceptual processing (and limitations) and the decision criterion found in classical signal detection theory (Green and Swets, 1966), is a direct consequence of the formulation of the optimization problem. As we shall see momentarily, the estimator portion of the OCM has already been used in models for monitoring, detection and decision-making. Because of its central importance, it is worthwhile to review briefly the operation of this estimator.

There are several ways to describe the operation of the Kalman filter (Gelb, et al., 1974) which serves as the estimator in the OCM. In the context of modelling human information processing, however, the description implied by Figure 2 seems most appropriate.[4] A "running" estimate, $\hat{x}(t)$, of the state-vector and the covariance matrix, $\Sigma(t)$, of the error in that estimate is maintained in "memory" along with an internal or "mental" model. The model, denoted by M, includes representations of: the system dynamics; the disturbance statistics; the observation noise statistics; and the known deterministic inputs. The matrix $\Sigma(t)$ may be thought of as a subjective confidence in the accuracy of the estimate. Based on this information, the operator is assumed to form, or predict, a set of expectancies, or prior estimates, concerning the *future* evolution of the state. This process can proceed open-loop (i.e., without observations) but in that circumstance, the operator's confidence in the estimate is likely to degrade with time. When observations are made, they are compared with the operator's prior expectation concerning that measurement (which can be derived from the prior estimate of the state and the portion of M that describes how states and outputs are related); the difference between these quantities is called the "residual", $r(t)$. The residual is then used to perform a "Bayesian" correction of the prior estimate, yielding a posterior estimate, along with a revised covariance, which are used to "update" memory and are available for control or decision-making. The correction to the prior estimate is proportional to the size of the residual; the proportionality factor is called the Kalman- or filter-gain. This gain is, roughly speaking, inversely proportional to the confidence in the a priori estimate and inversely proportional to the confidence in the measurement (i.e., to the observation noise covariance). Thus, the more confident the filter is of its prior estimate, the less it will weight the new data in correcting that estimate. And, for a given a priori

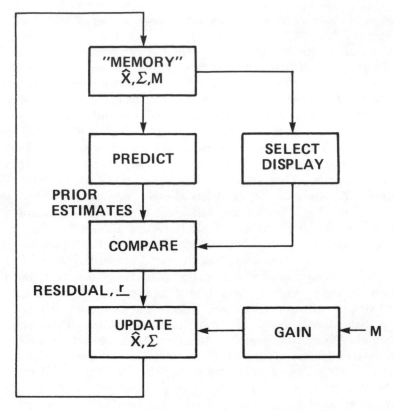

Figure 2. Model Structure for Estimator/Predictor.

confidence level, the better the data (i.e., the lower the observation noise), the more it will be weighted in correcting the prior estimate.[5]

When the assumptions underlying the Kalman estimator are satisfied (linear systems, gaussian random processes and an internal model that is "matched" identically to the true system), the estimator yields a minimum variance estimate and the residuals are a white noise process (with zero-mean and known covariance). In other words, the estimator extracts from the observations all the information that is available about the system states. Furthermore, the estimate and the covariance matrix are sufficient to specify the conditional probability density of the state (conditioned on the measurements) and they constitute a sufficient statistic for testing hypotheses about $x(t)$ based on the perceived data, $y_p(s)$, $s \leq t$. These features are very important in using the estimator for decision making. Notice that the system and the noise statistics can be time-varying without altering these properties. It should also be mentioned that the estimator is optimal for a given observation sequence or strategy, regardless of whether or not that strategy is itself optimal in the sense of minimizing error.

B.   Models for Monitoring and Decision-Making

Several different monitoring and decision-making models have been derived using the optimal estimator and the models for human perceptual limitations that are incorporated in the OCM. These models have been focused largely on the detection of failures or "abnormal" events, which are characterized either by system variables exeeeding prescribed tolerances or by the systematic deviation of system outputs from their expected behavior.[6] It is apparent that the optimal estimator generates outputs ($\hat{x}(t)$ and/or $r(t)$) that could be used for detecting either situation. Further, the estimator outputs may be used to derive either descriptive or prescriptive models of failure detection. For example, a descriptive model based on the state-estimate itself could involve decision regions in the state space. Detection of when the state is in a given region could be modelled as a deterministic decision based on the estimate alone, or it could be a probabilistic assessment based on the probability density function. A descriptive detection model using the residuals could be based on a test of the "whiteness" of the residual process with significant deviations from the ideal indicating an "event." For prescriptive models of decision-making, optimal decisions must be generated. This, of course, involves specification of decision criteria.

The first use of the OCM information processing structure in modelling such a human decision-making process was by Levison and Tanner (1971). They studied the problem of how well subjects could determine whether a signal embedded in added noise was within specified tolerances (a continuous, visual analog of classical signal detection experiments). From a practical standpoint, the decision is analogous to the problem of deciding whether conditions warrant a given action; e.g., deciding whether one is in the approach window on landing or deciding whether tracking errors are small enough for weapons release. They modelled this situation by assuming that the operator is an optimal decision maker in the sense of maximizing expected utility. For equal penalties on missed detections and false alarms, this rule reduces to one of minimizing expected decision error. The decision rule is simply a likelihood ratio test that, effectively, uses the densities generated by optimal estimator.

The predictions of the Levison-Tanner decision-making model compared favorably with experimental data for a variety of conditions involving different signal/noise ratios and different noise bandwidths. In addition, the model was tested for the case of two concurrent decision tasks and a concurrent decision and manual control task. This provided a broader test for the task interference model, used previously only for continuous control tasks (Levison, Elkind and Ward, 1971). Good results were obtained for the two decision tasks but not for the combined control/decision task. A number of methodological explanations for the lack of better agreement in the latter case were suggested, but the validity of the tested interference model for such combined control/decision tasks remains an area for further investigation.

The idea of using the residual sequence generated by the estimator structure of the OCM to model human failure detection was suggested by Phatak and Kleinman (1972). They also showed how one could employ "multiple" internal models with multiple estimators to perform simultaneous detection and identification of a failure, assuming that the failure corresponded to one of the hypothesized internal models. The work of Phatak and Kleinman was theoretical and did not contain validation data. However, other studies using "monitoring" of the residuals for detection of failures have tested some of these ideas.

Gai and Curry (1976, 1977) have used the OCM information processing structure to analyze failure detection in a simple laboratory task and in an experiment simulating pilot monitoring of an automatic approach. In both cases, a step or ramp was added to an observed signal at a random time to simulate a failure. This produced a non-zero mean value for the signal and for the residual; failure detection consisted of testing an hypothesis concerning the mean of the distribution of the residuals.[7] Sequential analysis was used to perform the hypothesis test. By summing (or integrating) the residuals, a likelihood ratio can be calculated and used to arrive at the decision. Gai and Curry modified classical sequential analysis to account for the fact that a failure detection problem is characterized by a transition from one mode of operation to another at a random time and the classical analysis is based on the assumption that the same mode of operation exists during the entire observation interval. They reported good agreement between predicted and observed detection times for both the simple and more realistic situations. In the latter case, the model was used in a multi-instrument monitoring task and accounted for attention sharing (via Levison's model) and cross-checking of instruments to confirm a failure. A significant result of the experiments was that the property of integration of the residuals appeared to be confirmed for both step and ramp type failures.

Wewerinke (1981) has extended the residual monitoring model of human operator failure detection and has performed a much more extensive experimental test of its validity. He also retained the information processing portion of the OCM, including the attention-sharing model. The decision-making or detection part of the model computes a short-term estimate of the mean of the residual sequence and uses this to compute the likelihood ratio which is then compared to a threshold to test for abnormal operation The decision threshold is related to (assumed) subjective values for miss probability and false alarm probability, as was also true in the work of Gai and Curry (1976).

Wewerinke performed two validation studies of the model. In the first, he used data from the basic laboratory experiment of Gai and Curry (1976). Decision error probabilities were fixed at .05 and the observation noise/signal ratio was set at the standard OCM value of .01 (-20 dB). The only remaining model parameter was the short-term averaging time for estimating the mean of the residual sequence. A value of four seconds for this parameter yielded an excellent fit to the data across experimental conditions; as noted by Wewerinke, this value is consistent with estimates of short term memory for visual stimuli.

In the second validation study, a variety of multivariable failure detection tasks were examined. Sixteen experimental conditions were investigated involving signal bandwidths, number of displays, mutual correlation of displayed signals and failure characteristics. Model parameters were fixed at the same values as for the first validation study. In the multi-display cases, the attention-sharing or task interference model was used with an optimal allocation of attention assumed; for this case the model was used to predict scanning behavior as well as detection performance. Model results were again in very good overall agreement with the experimental results both in terms of predicted detection times and scanning behavior. In addition, the specific effects of task variables on the experimental measures were predicted accurately by the model.

A recent model developed by Pattipati, Kleinman and Ephrath (1979, 1980) further extends the control-theoretic approach to modelling human decision-making. This model, called the Dynamic Decision Model or DDM, addresses the problem of task selection in a dynamic multi-task environment. An experimental paradigm of Tulga (1979) was modified to provide an appropriate idealized laboratory task for validating the DDM. The task involved observing separate rectangular bars moving with different velocities across a CRT. The height of each bar represented its reward or value. Dots were shown on each bar to indicate the time required to process the bar, that is, to gain its reward. The "processing" of a bar (task) was accomplished by holding down a corresponding push button for the required time. At any time, there were, at most, five bars on the screen.

As with the other decision-making models discussed thus far, the DDM used the basic information processing model of the OCM. Because of the independence of the tasks, the estimator decoupled into N-independent Kalman filters. Pattipati et al. introduced the notion of a decision state as distinguished from a task state. The decision-states are the variables that are oriented towards the decision action. They are a memoryless, nonlinear transformation of the task states and are analogous to, but somewhat narrower in concept than, the general notion of a "situation" that is defined later in Section III. In the paradigm modelled with the DDM, the decision states for each task are the time required and time available for completing that task.

In the DDM, the decision strategy is assumed to be completely myopic; which is to say, possible future actions and rewards are disregarded in the decision policy of the human.[8] (The measure which is to be maximized in a deterministic decision process is completely analogous to the expected net gain used in the DEMON model discussed below.) This assumption may be viewed as a constraint on human information processing, as an unconstrained optimal policy would consider the complete future courses of action in choosing the current one. Humans are certainly capable of considering, to a degree, future actions and consequences (planning!) so the completely myopic strategy is perhaps overly constraining. On the other hand, its selection removes a parameter from the model (the look-ahead time), simplifies computation and, in the case of the DDM, leads to results that agree with the data.

The DDM differs from the other decision-making models discussed in this paper in that human randomness is introduced in the decision-making algorithm itself. This is accomplished by assuming a distribution for the payoff values (attractiveness measures) and then incorporating Luce's stochastic choice algorithm (Luce and Galanter, 1963). Though the practical value of including this randomness in performing system design and analysis may be argued, it is clear that such randomness in human decision-making does exist. Moreover, in the context of the relatively simple paradigm to which the DDM was applied, the introduction of the stochastic chioce axiom allows the DDM to be used to compute performance statistics analytically, rather than by Monte Carlo simulation. Of course, significant deviations from the simplifying assumptions used in the basic DDM validation task could compromise the possibilities for analytic solution.

The DDM can be used to predict a number of metrics related to decision-making performance in the task-sequencing situation, including task completion probability, error probability (where error corresponds to starting a task that can't be completed), etc. Pattipati et al. report good agreement between model predictions (using fixed, OCM-determined values for observation noise and time delay) and experimental data across a range of conditions.

In addition to providing useful models of selected monitoring and decision-making tasks, the above studies provide further, independent, validation of the display processing and state estimation models developed for the OCM. When these results are added to the implicit validation provided by the tracking data, one has a strong case for this type of modelling of human information processing. The studies also extend the notion of situation assessment, in the control-theoretic modelling context, to include the detection of state-related events. These events may be defined in terms of "regions" in the multidimensional state-space or in a suitable output-space. Or, they may be events that signal a deviation from expectancy which can be detected by monitoring and testing the residuals. The detection of these events will, in general, trigger subsequent actions (such as selecting what to do next as in the DDM). Because of the commonality of the display and information processing aspects of the models, we may view these models as expansions of the OCM in the manner illustrated in Figure 3.

C.   Multi-Task Supervisory Control Models

The models just discussed are significant steps in developing control-theoretic models for supervisory control. However, these single task[9] or "single decision" models have not incorporated a number of very significant features common to many more realistic supervisory control problems. For example, none of the models consider any of the following: multiple tasks having different objectives; the detection of events not explicitly related to the system state variables; or multi-operator situations and the effects of communication among such operators. Perhaps the chief

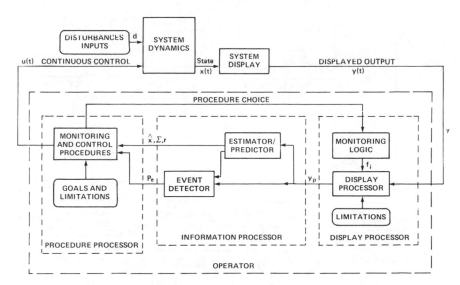

Figure 3.   Model for Continuous Control and Event Detection.

shortcoming of the models with respect to realistic supervisory control tasks is that they do not include the procedural activities of the operators or the discrete tasks that are often part of such procedures. Interestingly, these neglected features are often the prime concern of psychologically-oriented models. We now describe briefly three supervisory control models for more realistic tasks that begin to address these issues within the framework of the control-theoretic approach.[10] It should be noted that while these models represent conceptual and theoretical advances, they have not yet been tested against experimental data.

1. DEMON

DEMON (Muralidharan and Baron, 1979, 1980) is a decision, monitoring and control model for analyzing the en route control of remotely piloted vehicles (RPV's). A block diagram of DEMON, from Muralidharan and Baron (1980) is given in Figure 4. Very briefly, the en route operator's task is to monitor the trajectories and estimated times of arrival (ETAs) of N vehicles, to decide if the lateral deviation from the desired preprogrammed flight path or the ETA error of any of them exceeds some tolerance threshold and to correct the paths of those that deviate excessively by issuing appropriate control commands ("patches"). Path de-

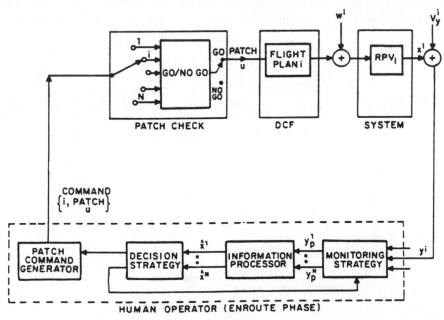

Figure 4.   DEMON Model for RPV Control Task.

viations arise from navigation errors and disturbances. In addition, the operator must decide when to "pop-up" or "hand-off" the RPV's under his control.

Display information is assumed to be updated at discrete times. It is also assumed that ETA and lateral deviation errors are presented separately and that only a single RPV can be selected for observation at a given time. Prior to a frame update, then, the RPV en route operator with N RPVs under control must also decide which among 2N + 1 displays to monitor. (An additional "display" is included to account for "secondary" tasks not included explicitly in the model).

The information processor in Figure 4 contains N Kalman filters to estimate the system state (i.e., the states of each RPV) and the uncertainty in that estimate, as in the OCM. With this information, the processor can also compute the subjective probability of exceeding various error tolerances or the proximity to a waypoint (or desired goal). Thus, the information processor provides an assessment of the situation or a "mental image" of the traffic picture.

The decision strategy generates both the monitoring and control choices. The operator's choices are assumed to be rational decisions governed by his knowledge of the situation, his goals and priorities and his instructions. These factors are incorporated in expected net gain criteria for monitoring and control.

The expected net gain (ENG) from a particular action is obtained by subtracting the cost of that action from its expected gain. The expected gain is the difference between the cost of events when no action is taken and the expected cost of events that may arise after this action. More formally, let a_i, $i = 1,2,...,n$ denote the alternatives open to the decision-maker at a given time and let a_o denote the "status quo" with respect to these alternatives. Then, for each a_i, the $ENG(a_i)$ is computed as follows:

$$ENG(a_i) = EC(a_o) - EC(a_i) - C(a_i)$$

where $EC(a_i)$ is the expected cost if alternative a_i is selected, $EC(a_o)$ is the expected cost if "status quo" is maintained and $C(a_i)$ is the action cost of implementing a_i. The rational choice is to select the alternative with the maximum ENG. Of course, specific cost expressions will depend on the tasks being performed. In addition, in computing these costs, some "decision horizon" or look-ahead period must be assumed. In DEMON, the frame update rate and the time required for pop-ups or hand-offs provided a time-scale for look ahead; in particular, a two-frame decision horizon was used.

The essence of the DEMON approach, as in other control-theoretic models, is to characterize the operator limitations and the mission goals and criteria for good performance in a manner that allows one to predict operator strategies and overall system performance. Operator limitations of observation noise, time delay and response bandwidth could be neglected in DEMON because they were insignificant in comparison to sensor measurement noise, display update rates and the time required for the discrete control inputs. However, a parameter was introduced into the information processing structure to account for the rate at which the operator's uncertainty about the vehicle state grows with time in the absence of further observation. This parameter relates to the operator's expectations concerning the disturbances perturbing the path of the RPV (as opposed to the true disturbances), as might be determined from instructions or through training. The mission goals were incorporated in the model in expressions for the ENGs for monitoring and control. They included factors such as thresholds for allowable errors and costs for monitoring and patching (control). These factors reflected priorities among RPVs as determined both by RPV type and mission time.

Model results were obtained illustrating the sensitivity of the performance predicted by DEMON to changes in parameters of the system (the number of RPVs to be controlled and the magnitudes of navigation and reporting errors) and in those describing operator behavior. These results show that the model behaves reasonably, that the model parameters do significantly affect performance and that the monitoring and patching trends are as expected. For example, results from DEMON predicting how performance in controlling the RPVs varies with the number of RPVs under control, suggest that four vehicles can be maintained within tolerance by a single operator. Errors can exceed tolerances for five or six vehicles and a critical point exists around seven RPVs. This prediction is consistent with findings in the literature concerning the abilities of air traffic controllers.

The DEMON model extended previous control-based approaches to a multi-task environment involving essentially discrete control decisions. (Note that the task interference model of Levison, Elkind and Ward (1971) treated multiple continuous-control tasks.) For DEMON, control of each RPV represented a separate task, each with a payoff for maintaining errors within tolerance and for timely pop-ups and hand-offs. Inasmuch as only one RPV (and only one RPV-state) could be observed at any time, the DEMON operator had to rely on memory (as embodied in the state-estimates) and prediction to decide when to monitor or "serve" a particular RPV and control objective. Another important advance was the introduction of the expected net gain decision-making algorithm. This algorithm can be related to classical subjective expected utility criteria and other methods. It also provides a very general approach to developing decision-making criteria in multi-task situations.

2. PROCRU

PROCRU (Procedure-Oriented Crew Model) [Baron, Muralidharan, Lancraft and Zacharias, 1980], is a model that has been developed recently for analyzing flight crew procedures in a commercial ILS approach-to-landing. PROCRU demonstrates how some of the additional features required to model supervisory control may be added to control-theoretic models in a reasonably integrated fashion. In particular, PROCRU is a multi-operator model that incorporates both "by the book" procedures and more unconstrained control and monitoring behaviors. It models continuous tasks directly and also accounts for the effects of discrete control tasks and for the time to perform them. Thus, the model combines some of the features of psychologically-oriented models with those of control-theoretic models. The development of PROCRU has had a significant influence on our thinking concerning the modelling of supervisory control and, therefore, will be discussed in some detail.

PROCRU includes a system model and a model for each crew member, where the crew is assumed to be composed of a pilot flying (PF), a pilot not flying (PNF) and a second officer (SO). In the present implementation of PROCRU, the SO model does not include any information processing or decision-making components. Rather, the SO is modelled as a purely deterministic program that responds to events and generates requests. PF and PNF, on the other hand, are each represented by complex human operator models which have the same general form but differ in detail.

Briefly, PF and PNF are each assumed to have a set of "procedures" or tasks to perform. The procedures include both routines established "by the book" (such as checklists) and tasks to be performed in some "optimizing" fashion (such as flying the airplane). The particular task chosen at a given instant in time is the one perceived to have the highest expected gain for execution at that time. The gain is a function of mission priorities and of the perceived estimate of the state-of-the-world ("situation") at that instant. This estimate is based on monitoring of the instrument displays and/or the external visual scene and on auditory inputs from other crew members and air traffic control (ATC).

The basic structure of the PROCRU model for either PF or PNF is shown in Figure 5 (from Baron, Muralidharan, et al., 1980). The system and system displays of the generic model are broken out laterally to illustrate the system states relevant to the problem, and the display cues available to the crew. The monitor (or display processor) of the model is partitioned to separate visual and auditory cues. The information processor is expanded to include both discret (event detection) and continuous (state estimation) information processing. Finally, the procedure

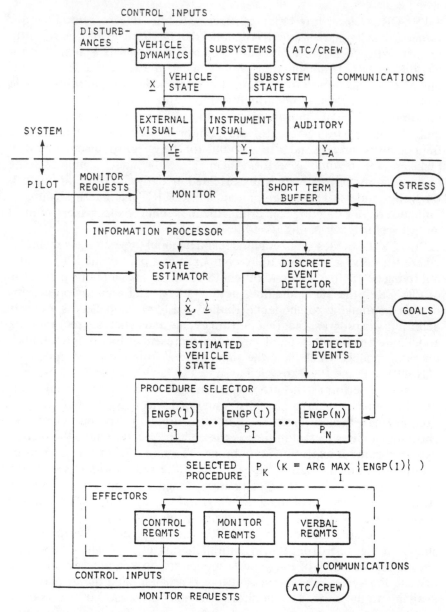

Figure 5. PROCRU Model Structure for Crew Member Analysis.

18

processor is separated into a procedure selector (which accounts for major decision-making in terms of procedure or task sequencing) and an effector block for implementing the actions called for by the procedures.

The monitoring and information processing portions of PROCRU are not unlike those of the OCM or other models discussed above, though they have some novel features and extensions (the reader is referred to Baron, Muralidharan et al. [1980] for a more detailed discussion). Two features are particularly noteworthy here: the estimator design and the treatment of auditory information.

The system dynamics incorporated in PROCRU are non-linear, so the basic linear Kalman filter could not be used. However, inasmuch as most maneuvers in the terminal area are standardized, it is possible to linearize the vehicle trajectory about nominal segments. Then, a linear estimator can be designed to estimate the perturbations from these "nominals." The estimate of the total state is then the sum of the estimates of the perturbation state and of the nominal state.[11]

Auditory information is treated in a different manner than is visual information in PROCRU. Briefly, we assume that auditory information is acquired instantly by the operator and that the correct message is stored in a short-term memory buffer (but is not "processed"). At the same time, the event detector is "notified" that a message is waiting to be processed. The subsequent processing of this message will, in general, depend on the nature of the message (alarm or communication) and on the time elapsed since its occurrence. If the message has a sufficiently high priority, the operator's activities will be interrupted and the auditory display will be selected for processing. The information in the buffer may also disappear or degrade in reliability with time, depending on the nature of the message. This treatment has the advantage that alarms are priority interrupts, but that they may be missed or "unattended to" during times of high activity or workload stress (e.g., when there are many alarms).

As noted previously, the operator is assumed to have a number of procedures or tasks that may be performed at each instant. The definition of these procedures is an essential step in the formulation of PROCRU. All crew actions, except for the decision as to which procedure to execute, are determined by the procedures. We emphasize that we use the term procedure here to apply to tasks in general; a procedure in these terms could have considerably more cognitive content than might normally be considered to be the case. Table 1 categorizes the flight procedures for the PF and PNF in approach to landing that are incorporated in PROCRU. For each crewman, six categories are shown, and for each category, specific types of procedures are itemized. These categories and types are discussed in detail in Baron, Muralidharan et al. (1980).

It is assumed that the PROCRU operator knows what is to be done

Table 1.   Procedures for PF (Pilot Flying) and PNF (Pilot Not Flying)
in PROCRU.

PF	PNF
VEHICLE CONTROL/MONITOR	
MANEUVER	VEHICLE STATUS DETERMINATION
REGULATE	FAILURE DETECTION AND
RETRIM	IDENTIFICATION
REQUEST/CALLOUT	
FLAP REQUEST	VEHICLE POSITION CALLOUT
GEAR REQUEST	ALTITUDE CALLOUT
CHECKLIST INITIATE REQUEST	APPROACH STABILITY CALLOUT
SUBSYSTEM	
ALTITUDE ALERT MONITOR/CONTROL	FLAP MONITOR/CONTROL
MISC. SYBSYSTEM MONITOR/CONTROL	GEAR MONITOR/CONTROL
	MISC. SUBSYSTEM MONITOR/CONTROL
ACKNOWLEDGMENT	
CHECKLIST ITEM	CHECKLIST ITEM
	ATC REQUEST
SAP/MAP	SAP/MAP
MISCELLANEOUS	
GENERAL MESSAGE RROCESSING	GENERAL MESSAGE PROCESSING
LANDING PARAMETER SELECTION	

and, essentially, how to accomplish the objective. However, he must decide what procedure to do next. This is a decision among alternatives and the procedure selected is assumed to be the one with the highest expected gain for execution at that time.

In PROCRU, the expected gain for executing a procedure is a function that is selected to reflect the urgency or priority of that procedure as well as its intrinsic "value." For procedures that are triggered by the operator's internal assessment of a condition related to the vehicle state-vector, the expected gain functions are appropriate subjective probabilities, as determined by the state estimation and event detection portions of the model. Procedures that are triggered by events external to the operator, such as ATC commands, communications from the crew, etc., are characterized by expected gains that are explicit functions of time. For either type of function, the gain for performing a procedure will increase, subsequent to the perception of the triggering event, until the procedure is performed or until a time such that the procedure is assumed to be "missed" or no longer appropriate for execution. The rate of increase of the expected gain function depends on the relative urgency of executing the procedure subsequent to the triggering event. This allows for distinguishing between procedures requiring immediate or fast action and those for which there is more latitude in the time of execution.

The model for procedure selection captures many important aspects of human performance in a multi-task environment, and is directly relevant to investigating the efficacy of flight crew procedures. It allows for procedures to be missed and/or interrupted: even flying the airplane may be neglected, as can happen. Although sub-procedural steps will not be performed out of order with this modelling approach, it is possible to pre-program such errors if desired.

The selection and execution of a procedure will result in an action or a sequence of actions. Three types of actions are considered: control actions, monitoring requests and communications. The control actions include continuous manual flight control inputs to the aircraft and discrete control settings (switches, flap settings, etc.). Monitoring requests result from procedural requirements for specific information and, therefore, raise the attention allocated to the particular information source. (Note that verifying that a variable is within limits may not require an actual instrument check, if the operator already has a "confident" internal estimate of that variable.) Communications are verbal requests or responses as demanded by a procedure. They include callouts, requests or commands, and communications to the ATC. Verbal communication is modelled directly as the transfer of either state, command or event information.

Associated with each discrete procedural action is a time to complete the required action. (It is possible to modify PROCRU to allow for a probabilistic distribution of action times.) When the operator decides to execute a specific procedure, it is assumed that he is "locked in" to the appropriate mode for a specified time (the "lock-up" time). For example, if the procedure requires checking a particular instrument and it is assumed that it takes T seconds to accomplish the check, then the "monitor" will not attend to other information for that period, nor will another procedure be executed.

In PROCRU, procedural implementation is modelled as essentially error free. However, errors in execution of procedures can occur because of improper decisions that result from a lack of information (quantity or quality) due to perceptual, procedural and workload limitations. If the effects of action errors are also to be analyzed, this is accomplished by deliberately inserting such errors directly into the model.

PROCRU generates a number of outputs that are useful for analyzing crew performance and workload. First, one can obtain full trajectory information. In addition to this information, one can obtain each crew member's estimate of the state and the standard deviation of the estimation error, the attentional allocation at that time and PF's control inputs. These data, along with significant events, etc., are tabulated in a file.

In addition to the trajectory output, PROCRU produces activity time lines. It should be emphasized that the time lines generated by PROCRU are closed-loop time lines; unlike those normally used in human factors analyses. That is, actions are not completely preprogrammed but depend on previous responses, disturbances, etc. Thus, one can change a system or human parameter in PROCRU and automatically generate a new, different time line. This ability to use PROCRU to investigate the effects of various changes on crew activity timelines (and on performance) is illustrated in the following discussion.

PROCRU simulations were made for two approach scenarios. One scenario corresponded to an evenly-paced approach with a long final segment to permit only approach stabilization. This "nominal" approach provided the conditions upon which the standard procedures were based. The second, "high-workload," scenario involved an increased tempo, resulting from an ATC request for an early turn towards the glide slope and, hence, a shortened final segment. The PROCRU simulations suggest that this simple, and common, change in ATC vectoring can have an interesting impact on the cockpit activity. It should be emphasized that the only input change made to the model to generate the "high workload" timeline was in the ATC command sequence; no changes were made in the crew procedures or in any other model parameters.

For a full discussion of the differences between the two approaches, the reader is referred to Baron et al. (1980). Here, we focus on the portion

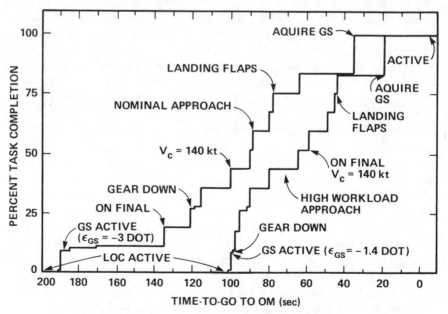

Figure 6. Non-Flying Task Completion Histories Predicted by PROCRU.

of the approach from the time the localizer becomes active to the time the aircraft passes over the outer marker. We also limit consideration to the completion of non-flying tasks by the crew. Figure 6 illustrates the effect of approach scenario on task completion during that time, with significant flight milestones indicated. It can be seen clearly that the high workload approach results in a requirement for much more rapid task completion once the localizer becomes active.

An examination of the milestone time line produced by PROCRU (Baron et al., 1980) provides additional detail concerning the increase in the pace of crew activity. In the high workload approach, the turn to final begins relatively close in to the field. Because of this, and the geometry of the ILS system, the glide slope (GS) becomes active almost immediately after the localizer. Moreover, instead of becoming active at a nominal -3 dots, the initial glide path indication is a relatively small -1.4 dot error.

The small GS error immediately places the crew "behind" in the approach, since the gear should have been dropped at 2 dots low, and the vehicle slowed to 140 kt at 1.5 dots low. Thus, there is an immediate gear down request by the PF. This is not acted on immediately because the PNF is busy announcing GS activation activity. To compound matters, the PF cannot initiate the required deceleration to 140 kt until he completes the turn to final. By this time, the GS error has been reduced to less than 0.5 dot, which means that the PF is "behind" in both his required deceleration to the approach speed of 139 kt, and his pitch down to the final glide path. Therefore, he "stacks" both of these commands immediately after ending the turn to final, in an attempt to catch up.

The increased tempo and late completion (as compared to nominal) of various tasks continues essentially until the outer marker (OM) beacon is crossed. At this point the crew has effectively "caught up" with the nominal profile and the remainder of the approach proceeds in virtually the nominal manner.

It may be argued, with some validity, that crews experienced in flying approaches of the high workload type would "plan ahead" and reschedule certain activities so as to avoid some of the problems caused by the modified vectoring sequence. This illustrates that PROCRU, as currently implemented, has only a very limited planning activity. The nominal timeline and associated procedures provide a (loosely-defined) implicit plan or script (Rouse, 1982) for executing the approach. The scheduling of specific activities along that overall plan is governed by the priority structures embedded in the gain functions and is adaptive to small changes in the nominal. However, to provide for the kind of rescheduling that would avoid some of the backlog seen in the high workload scenario, it would be necessary to modify PROCRU.

3. AAACRU

The approach used in PROCRU has also been employed to develop a supervisory control model called AAACRU for the commander/gunner crew of an anti-aircraft artillery (AAA) system (Zacharias, Baron and Muralidharan, 1982). Here, we consider only the model for the commander. The general structure for this model is essentially identical to the PROCRU model structure of Figure 5, except that an explicit situation assessor has been added to the information processing function of the model.

The commander's responsibilities in the AAA system include: monitoring system performance so as to be able to make appropriate situational assessments concerning possible counter-measures (CM) usage by the target; making decisions concerning the appropriateness of the current system operating mode (and the gains to be had by switching modes); making decisions concerning firing strategy; and communicating firing and mode switch commands to the gunner. In the model, a situation assessment logic computed CM probabilities and engagement progress, which "fed" a decision-making logic to maximize expected engagement kill probability based on an internal performance matrix model.

For this discussion, we focus on the information processor for the commander model. The information processor portion of the model consists of three sub-models, an "estimator," a "discrete event detector" and a situation assessor.

A hybrid linear/non-linear target state estimator was built from a "core" estimator which operated with (Cartesian) linear dynamics and a measurement interface which operated with non-linear spherical measurements. Although more complex than conventional linear (Kalman) state estimators which have been used in past AAA studies, it eliminated the need for postulating *ad hoc* internal models of the dynamics associated with the spherical coordinate target kinematics. The approach also provides an estimate of Cartesian-frame target acceleration, which is critical to situation assessment. Outputs of this estimator were input directly to the situation assessor and/or the discrete event detector. Outputs were also used to provide additional estimates of relevant information concerning engagement progress, such as range-to-target, time-to-crossover, etc. (which are analogous to the decision states of the DDM).

The information processor also contains an element that detects (or estimates the probabilities of) discrete events that are germane to the task, such as a CM onset (that did or did not result in an alarm), a verbal input from the gunner, or some annunciated condition. The inputs to the event detector are outputs of visual alarms, auditory information, *and* the outputs of the state estimator. The state information is used to detect state related events, such as a target maneuver (a tactical counter measure—TCM).

The situation assessor model takes in the estimated target state variables and the detected events (or their occurrence probabilities), and generates situational probabilities associated with CM usage and engagement progress. The situation assessment logic addresses five basic questions concerning the target that are of interest for the commander: (1) Is the target maneuvering (TCM)?; (2) Is the target using electronic counter measures (ECM)?; (3) Is the target using optical counter measures (OCM)?; (4) Is the target within firing range?; and (5) Is the target near crossover (point of closest approach and maximum relative velocity)? Questions 1 and 4 involve comparison of the estimate of components of the vehicle state with some predetermined set-points. However, certain discrete events (e.g., wing "glint") also affect the assessment of TCM and are combined with the state-related events in a Bayesian estimator. Question 5 involves computation of the time to reach a "deadline" and is similar to the time-available decision state of the DDM. Questions 2 and 3 involve detection of non-state-related events. The probabilistic assessments of these important engagement factors are incorporated in a five-component "situation vector," S, which is then used in determining subsequent decisions to be made by the commander.[12]

III. A GENERIC MODEL FOR SUPERVISORY CONTROL

The previous section demonstrated the possibility of developing more complex and complete supervisory control models based largely on a control-theoretic viewpoint. Here, we present and discuss a generic conceptual model that reflects the experiences gained in reviewing and developing these models as well as those obtained in a recent study to examine the feasibility of applying the control-theoretic approach to modelling operators of nuclear power plants (Baron, Feehrer, Murali-dharan, Pew and Horwitz, 1982). As expected, this model shares many features of the models described above. However, it also includes conceptual extensions and generalizations that we believe will be desirable or necessary for future applications. The model is conceptual at this time in that some of its features have yet to be implemented, let alone tested, in any supervisory control model.

Figure 7 illustrates the general supervisory control model structure. The model is a closed-loop man-machine "simulation" that incorporates elements that can represent a range of operator behaviors: cognitive and psychomotor, and continuous and discrete. Below, the major elements are discussed with special emphasis on the information processor, because of its central importance to the cognitive aspects of supervisory

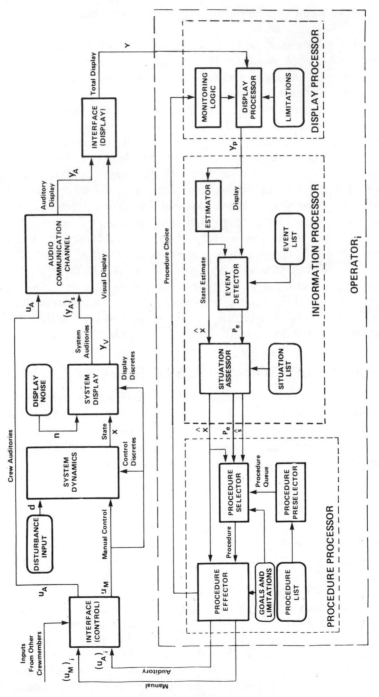

Figure 7. Generic Control-Theoretic Model for Supervisory Control.

26

control and because it is in this part of the model that new extensions are suggested.

A.   Mathematical Model of the System/Environment

It is necessary to include a mathematical description of the system/environment in the supervisory control simulation model. The degree of detail and the level of complexity of the system model will depend on the specifics of the issues to be addressed. However, in general, the model must include those factors that are needed to perform a closed-loop system analysis, as listed below:

- a state-variable description of the processes including any automatic control and engineered safety features
- a description of potential disturbances (known and unknown)
- a description of the instrumentation and display information provided for the crew (including information content, alarm set points, instrument or sensor noise, update rates and failure nodes)

The state variable description of the system will consist of differential (or difference) equations describing the time evolution of system variables. It is important to note that this system model can be nonlinear and/or time varying. In addition, it will frequently be necessary to simulate both normal and off/normal conditions. However, in many cases, it may not be required to model the fast response of the system accurately as these may be controlled automatically or may simply be of secondary interest. (For example, in PROCRU, where the prime concern was with flight management, inner-loop aircraft dynamics were ignored).

B.   Model for the Human Operator(s)

The lower half of Figure 7 illustrates the model for the operator. It is always difficult to "parse" human behavior, and one might well imagine other breakdowns than those shown and discussed below. Certainly, major activities like situation assessment transcend the boundaries indicated. Nonetheless, we believe that, on the whole, it is important for clarifying thinking about behavior to adopt some structure. The one we have selected here is a slight expansion of the structures used in Section 2 and emphasizes the development trend on which we have been focusing.

1.   Display Processor

The display processor portion of the model has two functions: it implements the "conscious" observation decisions of the operator by selecting the appropriate displayed quantity; and, it accounts for sensory and processing limitations associated with observation.

We assume that the operator is a single-channel processor of information. In terms of observing displayed information, we will assume that only one information "display" can be "attended to" at any time. The selection of a particular source of information is governed by goal-oriented processes: thus, it will depend on the purposes for which information is being gathered. Much of the time, the choice of a particular source will be procedurally driven; i.e., the execution of specific operating procedures will direct the choice. However, for an activity that is not governed by well-defined procedures, the display selection process incorporated in the model will be designed to support that activity. For example, monitoring during normal conditions or during diagnosis of a problem by other than procedural means (see below) will involve normative or heuristic "scanning" models that drive the display selection process.

As illustrated in Figure 7, the operator processes both visual and auditory information[13] coming from the various displays and the communication channels. In general, the information coming to the display processor is of two types, regardless of the source: quantitative (continuous-valued) information concerning process variables (e.g., temperature, and pressure or position and speed), or discrete information concerning process or equipment status (e.g., temperature $> T_{limit}$, or valve A closed, or landing gear down).

To model the operator's observation of a quantitative visual display, the basic "perceptual" model contained within the OCM, suitably modified, is used. Thus, observation noise is included as a limitation but, for many problems, the human's information processing delay of approximately .2 seconds is subsumed in a longer display fixation or "lock-up" time.

Discrete displays that provide indications of status are considered as event indicators, which, when "attended to" directly, subserve the operator's functions of state estimation/prediction, event detection and situation assessment. The discrete displays will generally not have an observation noise associated with them, but they can have corresponding subjective or objective reliabilities. Many discrete displays are auditory in nature and, in this case, they are treated in the same manner as in PROCRU (see above).

2. Information Processor

The information processor is comprised of three elements: an estimator/predictor, an event detector, and a situation assessor.

The estimator/predictor performs two functions. First, it processes the observed information to update its estimate of the process state variables. And, second, it predicts the future evolution of the system on the basis of the estimate of the current state, known inputs to the process, and a "mental model" of the plant.

As we have seen, the use of a Kalman filter to model human estimation behavior has proven to be successful for a number of control and decision/detection tasks. However, problems arise in employing it directly in modelling human estimation in complex supervisory control problems; in particular, most often the system is non-linear and it is of such complexity and high order that the notion of a perfect mental model is difficult to accept.

There are several possible approaches to dealing with the system non-linearity. In some special cases, it may be possible to derive a nonlinear estimator directly, as was done in the AAACRU model. Most often, this will not be feasible and some form of linearization will be required in developing the estimator. In PROCRU, linearization about "nominal" trajectory segments was used to derive the estimator. These techniques were appropriate for the applications investigated but, in the author's view, the approach that is likely to prove most useful in the long run is extended Kalman filtering (Gelb, et al., 1974) which employs linearization about the latest estimate of the state. The theory for building such an estimator is well-established and its use has several additional virtues: (1) it preserves the fundamental predict/correct logic of the Kalman filter that is so appealing from the standpoint of representing human information processing; (2) it assumes that the operator does linear prediction so, at least in one sense, the mental model complexity is reduced; (3) it generates a set of filter residuals that can be used for diagnostic purposes (see below); and (4) it reduces to a well-validated model in the special cases where the linear Kalman filter is optimal.

Though the extended Kalman filter requires a model of the system, there is nothing in its logic or structure that necessitates that the model used to implement the filter be an exact replica of the true system or even a close approximation to it.[14] We shall assume, therefore, that the estimator employs the same logic as the extended Kalman filter but we place no a priori restrictions on the "mental" or internal model used by that estimator. Thus, the operator's mental model can differ from the system model and it can also change with time and with situation changes. For example, after a system failure, the mental model is expected to be "wrong" until the failure is diagnosed and the model is suitably modified. On the other hand, during normal operation, it is not unreasonable to expect the operator to have a good model of the system. Furthermore, if we wish to bound information processing performance by using a normative model, so as to explore system (rather than human) deficiencies, then the assumption that the operator's model replicates the system model is a useful one.

The basic difficulty with the above approach is that the operator's mental model must be defined explicitly and quantitatively and, once one departs from the normative approach, there exists a theoretically infinite number of possible choices for the mental model. There is, therefore, a potentially dramatic increase in the number of model parameters that are free. One might argue, however, that this generalization actually strengthens the approach, not only because it forces the modeller to make the assumptions concerning the mental model explicit but also because it allows one to explore the consequences of different operator conceptions about the system. Thus, for example, if different training approaches can be expected to lead to different mental representations of the system, then the effects of such training differences may be predicted by the model. Additional comments concerning the selection of a mental model along with the equations that can be used to implement the estimator are given in Appendix C of Baron, Feehrer et al., (1982).

The outputs of the estimator/predictor are the operator's *a priori* (before observation) and *a posteriori* (after update) estimates of the process state, the respective subjective estimation error covariances which, in general, will now differ from the true error covariances, and the residuals and their covariances.

We will assume that the operator's subjective probability distribution for x is normal with conditional mean $\hat{x}$ and covariance Σ. Thus, the estimator produces the status information that is often needed for control purposes and the subjective probability estimates that can be used for decision-making and/or detection.

The discrete event detector is intended to model those aspects of operator information processing, other than estimation and prediction, of the process state variables. Typically, it will be concerned with determining or detecting that an event has occurred that helps to define a situation, thus "enabling" a procedure selection and execution. The event may be a transient (that did or did not result in an alarm), a request for action (say, from another crew member), an alarmed condition (such as exceeding a threshold or a tripped component alarm), or the verification of the accomplishment of an intended action. The inputs to the event detector are visual alarms, auditory information, the outputs of the state estimator/predictor, and the list in memory of possible events. The state information is used to detect state-related events, such as exceeding desired limits or failure of the system to respond appropriately to a control input. The outputs of the event detector are the occurrence probabilities of the various events, P_e. The detected events are stored in memory where they can be used for situation assessment (see below).

It is anticipated that different detection processes would be used for different types of events. For initial detection of an auditory message, the presence of the message would first be detected with probability one.[15] The presence of this message would be made known to the situation assessor, and, if and when the message was subsequently processed (i.e., decoded), its contents would be available to the event detector. Conditions on process state variables would be evaluated (detected) on the basis of the state estimate and its uncertainty. The output of the event detector would then simply be P(x≤ SET-POINT).

More sophisticated event detection models are envisioned to support higher level diagnostic activities. These are likely to be drawn from existing models for human failure detection (Section II) or as modifications of other system failure detection algorithms (e.g., Willsky, 1976 or Chow, 1980; Caglayan, 1980), as appropriate. Virtually all of these techniques use filter residuals. Thus, we foresee a residual monitoring system that searches for deviations from expectancy by testing for biases in the residuals. The same approach may be used for confirming that the results of an action are as expected. The use of such models is consistent with the notion that, for an operator to perform the detection task in a satisfactory manner, he must share certain feature of "good" failure detection systems of an inanimate nature.

In general, we believe the human is capable of employing analogues of any or all of the above schemes, but that the different methods place different demands on his cognitive processing resources. Therefore, we envision the event detector as being able to draw on a range of such schemes or algorithms, depending on the situation.

In the supervisory control models described in Section II, the idea of situation assessment was extended from state-estimation to include state-related events. For many problems, particularly those involving diagnosis or higher level problem solving activities, there is a requirement to expand the notion of situation assessment still further.

We define a "situation" to be a single condition or a set of conditions on underlying system variables and defined events (not necessarily related to state variables) that is identifiable in an aggregated sense (i.e., can be given a name) and that serves to define a set of acceptable alternative responses. Simply stated, situations are the predicates that must be satisfied to "fire" the actions in a production rule (i.e., IF (SITUATION) THEN (ACTION) or IF (SITUATION) THEN (SITUATION)). In this terminology, such disparate conditions as a threat alert or a message waiting to be processed would qualify as "situations."

The situation assessor block of the supervisory control model is aimed at computing the probability or likelihood (or other decision function) of a postulated situation. Its inputs are the outputs of the information processor and an ordered list of possible situations that are stored in memory. The ordering of the situations is assumed to be based on prior estimates of the probability of occurrence (which can be updated as the situation evolves) and on the potential consequences associated with the situation.

Thus, threatening situations would be expected to head the list of situations to be considered.

As noted earlier, the situation may correspond to a single condition on a process variable or on equipment status. In such a case, situation assessment and event detection may coincide and the process is straightforward. More generally, a situation will be defined in relation to a larger set of conditions. However, in practice, the operator's assessment of the situation may not incorporate all the diagnostic elements. That is, although, in principle, the number and character of the symptoms or events required to identify a given situation uniquely are specifiable, limitations on the accuracies of sensor and display equipment, combined with limitations on the memory and information processing capabilities of the human, often result in (real-time) evaluation of less than the full complement of maximally situation diagnostic information. In such cases, and in those where given conditions cannot be evaluated with certainty, the resulting diagnosis is probabilistic in nature, embodying an implicit statement that this is "situation A" with some level of confidence. The model we propose for this process attempts to capture this and other aspects of operator situation assessment behavior.

First, we define the concept of a template.[16] A template is an explicit specification of the value required on each of n system parameters and/or variables in order to justify the diagnosis, at a given level of confidence, that a given situation obtains. Thus, a template is an aggregation of events. Note that the "situations" defined in the AAACRU model can be placed in this context. Typically, a template contains fewer diagnostic parameters than are potentially available, considering the aggregate of available information, and its utility as a problem-solving tool varies both with the number and diagnostic value of the parameters contained therein.

The sizes and contents of templates are largely defined by the diagnostic procedures contained in operations manuals and by the accumulated training and experience of the operator. Depending on length and complexity, all, or portions of given templates may be stored in memory and, during a given diagnostic episode, an operator may employ both written and remembered diagnostic procedures. As with situations, a list of templates is assumed to be stored in memory and ordered according to the perceived utility for resolving the situation considered (which would depend on objective and subjective factors such as training, experience, etc.).

After a coarse pre-filtering of situations to eliminate obvious untenable hypotheses (situations), detailed diagnosis at a finer level commences. The search through a template is assumed to proceed sequentially, one element or condition at a time. After each evaluation a decision is made (in the procedure selector) whether to accept or reject the hypothesized situation or to collect more data. If the situation is accepted, an appropriate response is chosen; if rejected, the next situation is considered. If the choice is to collect more data, three options are available: (1) further observations to resolve the template condition currently being considered; (2) testing of the next element in the template; and (3) abandoning the symptom-oriented matching for a more sophisticated event detection or hypothesis testing scheme.[17]

Thus, the situation assessment algorithm models search and diagnosis as a modified sequential testing procedure which initially uses symptoms as a diagnostic tool and moves to other, more sophisticated, means if the symptoms failed to resolve the problem. This model would be a system-theoretic parallel to the skill-rule-knowledge-based hierarchy of Rasmussen (1980a). It would also be analogous to that author's characterization of symptomatic-, topographic-, and hypothesis-and-test search strategy modes (Rasmussen, 1980b). The approach is also similar to Rouse's model for problem-solving (Rouse, 1982) which is not surprising, given that model's relation to Rasmussen's work. The overall situation assessment model is illustrated in the flow chart of Figure 8.

We close the discussion by noting that situation assessment is expected to effect the estimation process. In particular, if the situation deviates from expected, then it must be assumed that the information used in arriving at that prior expectation was "defective." This should increase the uncertainty about the estimate of the state and could cause a revision of the mental model. This is illustrated in Figure 9, where we show the situation assessment function as an outer-loop (or supervisory monitor) for the basic estimation process. It is important to note that there exists in the control theory literature (e.g., Caglayan, 1980) systematic methods for modifying the estimates and for increasing their uncertainty, if the data indicate that the original internal model is inadequate.

3. Procedure Processor

The "procedures" are the means by which the operators organize and carry out their monitoring, situation assessment and control responses so as to accomplish their objectives. Procedures exist in "manuals" or they reside in memory, having been learned through training and/or experience. We allow both highly structured formal procedures and other less structured, but goal directed, responses (which we will also call procedures, for convenience).

A formal procedure is a specific sequence of tasks or actions (perhaps of length one) together with the "situations" that trigger those actions. Each step in such a procedure is a production rule with the general form: IF (SITUATION) THEN (ACTION) or IF (SITUATION) THEN (SITUATION). As we have seen, the determination of whether or not a sit-

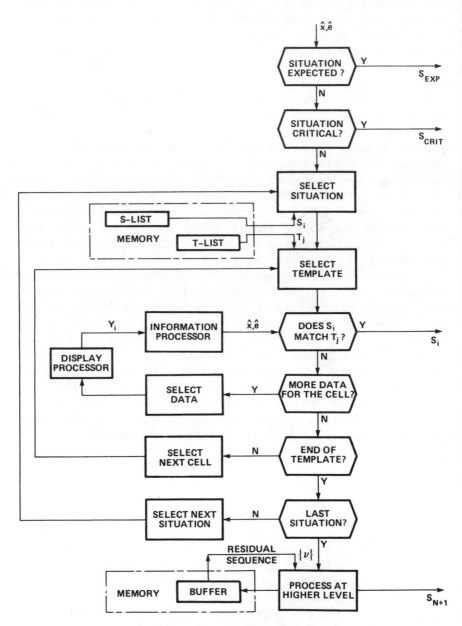

Figure 8. Flowchart for Situation Assessment.

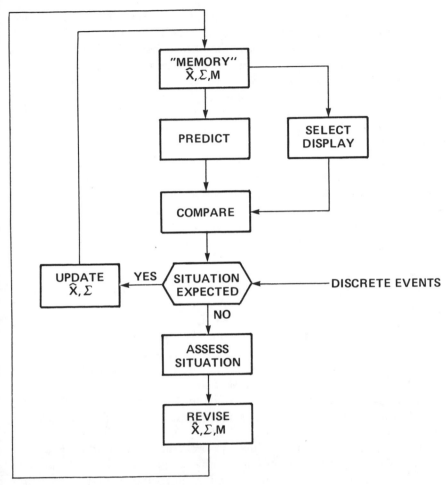

Figure 9. Modified Structure for Information Processor.

uation is true is generally probabilistic and is made in the situation assessor portion of the model.

Informal procedures are those that are to be formulated and employed when the existing situation does not "fire" a known formal procedure. They are included partly in an attempt to model some knowledge-based behaviors and partly for modelling convenience. Examples of informal procedures are: continuous control laws; monitoring strategies for detecting anomalies during normal (unalarmed) conditions; situation diagnosis algorithms when procedural diagnostics fail to resolve a situation; and message decoding (the process of "reading" the auditory buffer).

Major decision-making at several levels takes place in the procedure selector block. The first decision is concerned with situation assessment and will involve some form of sequential test that will be performed to decide whether to accept/reject or defer the identification of a situation as discussed above. The thresholds for this test might well depend on the situation being evaluated. Thus, for example, a plant threatening situation might have a lower threshold for acceptance than a non-threatening situation. In general, if the probability of a situation is known, the thresholds may be determined from decision analysis as illustrated in Baron, Feehrer et al. (1982).[18]

If the decision has been made to defer the identification of a situation, the modelled operator will have two options: collect more data using the current diagnostic test procedure (this could involve obtaining information on new variables or waiting for the situation to evolve); or, try a new diagnostic (situation assessment) algorithm. Thus, a second decision is required.

If the situation is identified, then, by assumption, so too are the next goal(s) and the options for achieving them. The situation may uniquely determine a goal and a procedure for accomplishing that goal; indeed, they may be identical so that the goal is, for example, do procedure A. If this is the case, and if the procedure is one requiring an immediate action, the operator is assumed to execute the procedure directly. If the procedure does not require an immediate action or, if it contains several steps, we assume that, at the end of each step, a decision must be made whether to continue, interrupt or terminate that procedure. Continuation would occur if the situation resulting from execution of the procedural step is consistent with the assessed situation. Interruption would occur as the result of a decision to suspend procedural execution because of the priority for processing an intervening alarm or because of the need or desirability of performing an alternative informal procedure (e.g., monitoring to verify instrument reliability). Finally, a decision to terminate would result from a change in the identification of the situation that resulted in the course of executing the procedure.

It is possible, though perhaps unusual, that a well-defined situation might admit or "enable" more than one possible formal procedure. In such a case, a formal decision among the alternatives would be required. A more likely occurrence is that alternative goals and/or options are possible for a given plant situation. However, in this case, too, the operator (model) would be required to make a decision among alternatives.

Thus, in all but the cases requiring immediate and pre-specified action, the operator is confronted with competing alternatives. The choice he makes is assumed to be rational, and would be computed in the model on the basis of an appropriate expected net gain (ENG) calculation, as was done in the supervisory control models described in Section II. The rational choice is to select that action which has the greatest ENG. In evaluating these alternatives, the (modelled) operator would have to "predict" the consequences of various actions using its mental model (or, perhaps a highly aggregated or simplified version of it).

The model structure we have adopted provides an opportunity to go beyond completely myopic strategies and to incorporate at least some aspects of planning. The planning process, according to Johannsen and Rouse (1979) may be broken down into the following aspects: (1) Generation of alternative plans; (2) Imagining the consequences; (3) Valuing of consequences; (4) Choosing and initiating plan; (5) Monitoring plan execution; (6) Debugging and updating plan. With respect to these facets of planning, only the *generation* of alternative plans is not considered to some degree in the envisioned model; instead, at least for the time being, alternative plans are assumed to be given.

The selection and execution of a procedure or other response results in an action or a sequence of actions. Three types of actions are possible: control, monitoring and communications. The control actions include any continuous manual control inputs to the system, and discrete control settings (set points, switches, etc.). Monitoring requests result from requirements for specific information; these requests are input to the display processor which selects the particular requested information source. We note that verifying that a variable is within limits may not require an actual instrument check, if the operator already has a "confident" internal estimate of that variable. Communications are verbal requests or responses resulting from interactions with other crew members. Mostly, they involve requesting or transmitting state or situation information.

Associated with each discrete action is a time to complete the required action. (It is possible to allow for a probabilistic distribution of action times). When the operator decides to execute a specific procedure, it is assumed that he is "locked in" to the appropriate mode for a specified time. For example, if the procedure requires "checking" a particular instrument and it is assumed that it takes t seconds to accomplish the check, then the display processor will not attend to other information for that period, nor will another procedure be executed. The one exception to this rule is that certain high priority auditory alarms can interrupt the activity. Note that the time required to obtain information would depend on the action required (eye movement, walking to another part of a plant, communication, etc.).

4. Memory

Although not shown explicitly in Figure 7, the various processes considered in the model require information (event and situation lists, templates, mental models, etc.) that is presumed to reside in the operator's memory. In our view, it is not necessary to model the storage and retrieval of information as a process, at least initially. Rather, memory can be viewed essentially as a "storehouse" of information that can be referred to, added to, or modified.

Limits on short-term memory of auditory information and other detected events are included in the model. In addition, "memory" of the state estimate will degrade with time following an observation. However, the time or "mental effort" required to retrieve information from memory have not been considered. It may be possible to estimate these quantities based on psychological considerations, but we suspect their impact on problems of interest to be of secondary importance, as compared to the short-term memory limitations that will be incorporated.

IV. SUMMARY AND CONCLUDING REMARKS

The basic objective of the control-theoretic approach to modelling supervisory control is to develop quantitative (i.e., mathematical/computational) models that will facilitate the investigation of the overall performance, reliability and safety of complex systems. The models we seek are aimed, principally, at questions that revolve around the interactions between system design factors and human limitations and capabilities in information processing, decision-making and control. As a result, the approach starts from the conviction that the supervisory control model must be a system model that includes dynamic representations of the systems to be controlled, the environment and the humans operating therein. Because the systems and problems of interest are so complex, it will generally not be possible to develop closed form analytic models, and time-based computer simulation models will be needed.[19] Thus, the models will be analogous to human-in-the-loop simulation.

The approach to development of the model(s) for the human operator(s) is top-down and starts from a description of the system that includes formalized statements of the goals and sub-goals for operation. The human operators are then considered as information processing and control/decision elements in a closed-loop. Thus, the attempt is to model the human in system terms at the task or functional level, as contrasted with bottom-up methods which attempt to synthesize human performance from a sequence of fundamental activities such as anatomical movements, memory recalls, etc. This reflects the belief that, for the large scale problems of interest, a more detailed bottom-up model of discrete activity is at best unnecessary and at worst impractical, a view that seems to be confirmed by the experiences with the HOS (Human Operator Simulator) bottom-up model, as reported by Lane, Strieb, Glenn and Wherry (1980).

The general thrust of the approach is to develop normative models (or sub-models) that prescribe operator performance with respect to the system goals (or sub-goals), but are constrained by operator limitations such as limited attentional capacity or imperfect knowledge. However, if normative solutions and models cannot be derived because of the complexity of the problem, as will often be the case, then sub-optimal models or ones employing goal-oriented heuristics are employed. The focus on a rational, but constrained, operator helps one to separate system-induced errors from purely human error. If behavioral deficiencies or blunders are to be examined, they are "programmed in," rather than predicted by the model. However, the consequences of these assumed behaviors can then be examined systematically.

The top-down approach facilitates the use of a modular model structure that allows for the incorporation of sub-task models that have been developed and validated in other contexts (such as those for failure detection described in Section II). This is important because experimental validation of the total system model may not be possible within reasonable time and economic constraints. Such a modular approach also reduces the requirement for new model development and it facilitates model evolution and modification (including software modifications).

Both continuous and discrete tasks are considered within this framework, but they are treated differently. For continuous tasks, operator limitations are modelled at the processing level (e.g., observation noise) and task completion times and execution errors are model outputs (i.e., predictions). For discrete tasks, completion times and errors (or distributions for these quantities) are inputs which must be estimated by the analyst using any available data base. During the execution of a discrete task, the operators' attention is assumed to be captured for the prespecified task completion time, unless interrupted by an event with a very high priority. Thus, discrete tasks are treated in a manner that is analogous to the treatment of sub-tasks in a network model but they are integrated into the overall control-theoretic approach. (The reverse procedure of embedding continuous tasks in a network has been employed by Miller and Seifert [1980]).

As is discussed in Section II, the general approach has been applied to manual control, decision-making, supervisory control of remotely piloted aircraft, the analysis of crew procedures in the cockpit of a commercial aircraft on approach to landing, and to examine the situation assessment and decision-making activities of the commander of an anti-aircraft artillery (AAA) system. These applications demonstrate a capability for modelling many of the activities associated with supervisory control and a basis for developing a generic model. Such a model has been suggested in Section III; this model expands on previous implementations, particularly in an attempt to consider the diagnostic activities of operators.

The generic model assumes that the operator is a single-channel pro-cessor of information and executor of actions. Though this assumption may be questioned, it is certainly a plausible one and it is most useful for bounding human performance. In addition, the model includes "lock-up times" for observation and action, where appropriate (e.g., eye-move-ment times or time to acquire a piece of data from a manual). These are times during which no other overt activities occur. These factors, along with the natural tempo of the task as simulated through the system model, can lead to model predictions of time stress and errors. They can also show where the design "induces" such effects on a highly skilled and knowledgeable operator who is constrained by fundamental human lim-itations.

As an example, a multiple alarm scenario is one in which we may expect the model to suggest or predict operator difficulties. Alarms are assumed to be "attention-getting" devices whose presence is known to the operator immediately through an auditory signal. However, the determination of which alarm is present may require a direction of visual attention to ac-quire the particular information. If there are multiple alarms, the modelled operator will have queue of high priority items demanding attention, a situation that could lead to delayed detection of important events or pos-sibly to missed alarms. Moreover, multiple alarms that occur simulta-neously, or nearly so, will complicate both the template-matching and the higher order diagnostic algorithms and could lead to long delays or "er-rors" in diagnosis.

The information processing portion of the model incorporates important extensions of previous models. It uses a basic predict/correct logic based on a mental model of the plant. The logic is analogous to that of the Kalman filter model used and validated in previous models for operator control and monitoring. However, assumptions on linearity and on the mental model have been eliminated. The logic provides a plausible al-gorithm for state-estimation and prediction. The mental models may be used for fast time prediction (or "thought experiments") to evaluate al-ternative options for action. Furthermore, the effects of different mental models (as might result from different training regimens or different roles) on overall performance may be analyzed.

A discrete event detector is also part of the information processor por-tion of the model. The envisioned event detector is more complex than those employed in previous models in that it is expected to employ a range of detection algorithms that vary with respect to computational sophistication. The simplest algorithms involve "attending to" and pro-cessing an out of tolerance condition, whether annunciated or not. The more complex algorithms employ the mental model of the plant to generate residuals which are tested for deviations from expected behavior.

The situation assessor portion of the proposed model is also an expansion of the previous models. As with the event detector, a heirarchy of algorithms is contemplated for situation assessment activities associated with diagnosis of a problem. At the lowest level is a template matching scheme which checks symptoms against a template that is part of a procedure residing either in a manual or in memory. The more sophisticated assessment algorithms would employ alternative mental models corresponding to the hypothesized situations and, by comparing the observed data with model outputs, would select the most likely hypothesis.

This model for the diagnostic process viewed *in toto* could exhibit or predict human errors known to occur under appropriate conditions. For example, under time stress, a diagnostic decision may be forced before evaluation of an adequate number of template elements has been accomplished—thus leading to an error. Or, at the opposite end of the spectrum, operator biases or uncertainties may lead to an unwillingness to abandon a given hypothesis or to make a decision without "more data," i.e., to cognitive lock-up.

In the model, the assessed situations are the "predicates" for actions spelled out in procedures. The procedures themselves range from highly structured rules to less structured, but goal directed, responses. If more than one situation is assessed as being possible, the selection of a procedure for execution is based on a rational "decision among alternatives" that takes into account the uncertainties in knowledge and the costs and payoffs associated with actions.

The envisioned supervisory control model could provide a number of interesting and useful outputs. Because it is a simulation model that includes the system, it can provide time histories of all system variables included in the model. Activity time lines showing operator actions would be available to determine periods of excessive activity, etc. These time lines would be significantly different from those traditionally employed in human factors analysis in that they would be generated in a closed-loop fashion. Operator behaviors and "errors" arising from the interaction of cognitive limitations and task-related problems (displays, procedures, etc.) would be predicted by the model. This output would include the performance effects of time stress. Internal (emotional or physiological) stressors would have to be modelled by modifying the basic model parameters related to the operator's processing limitations such as observation noise, lock-up times and the mental model.

In addition to the above model outputs that are fairly standard and are directly verifiable, there are several that could prove most useful but may be impossible to measure directly. For example, the model provides measures of the operator's current estimate of the state and the current contents of memory; if the multi-operator model is considered, the "collective" or aggregate values of these quantities is available.

A brief analysis of the data requirements for the model suggest that much of what is needed may be found in the literature. Because the operator model draws on previously validated sub-models for many tasks, there exists appropriate operator data for these tasks. Critical needs for model development in a particular context will exist in the area of specifying the goal/sub-goal hierarchies and values and in the definition of suitable "mental models" for the task. For the present, these requirements will probably have to be satisfied by pre-modelling task analysis and through operator interviews.

Naturally, given the complexity of supervisory control and the relative infancy of attempts to model operators in these tasks, there are a number of significant issues that are unresolved, partially treated, or currently neglected and are therefore areas requiring further research; we mention some of the more prominent problem areas below.

The proposed situation assessment/diagnosis model is, at present, a conceptual construct plus some potential algorithms. Though, as noted in Section III, the model may be viewed as a control-theoretic analog of other approaches to modeling human diagnostic or problem-solving behavior (Rasmussen, 1980a; Rouse, 1982), it has not been tested in a well-designed, constrained laboratory experiment, let alone in a realistic situation. It is clear that some such test is necessary before the model could be used with any confidence in its predictive validity.

As has been discussed throughout, the goals of the operators play a very significant role in determining the predicted assessments, decisions and control strategies of the modelled operator. The formulation of goals and sub-goals in an unstructured situation is a very complex process and modelling this aspect of behavior is, virtually, virgin territory. In more structured situations, such as those involving well-trained operators supervising systems of the type we have been discussing, it is more likely that goals can be characterized at a level of detail sufficient to permit modelling of goal-oriented operators. This could be accomplished via an analysis of information in operating manuals and through operator interviews. Even in such cases, however, the techniques for defining the goal-structure and the specific goals are not well-developed and require further work.

We have touched on the modelling of planning in various portions of the discussion. There is the capability in the model to go beyond completely myopic strategies and to incorporate the evaluation, selection and monitoring of given alternative plans. However, there is a basic difficulty in this approach, in that there is a requirement for prespecifying all alternatives in advance. Clearly, an algorithmic approach to the generation of new plans would be desirable but it is likely to be far more difficult to achieve. The modelling of planning activities of human supervisors is an area for both theoretical and empirical research.

In modelling multi-operator systems, the model attempts to account for the transfer of information between crew members and the communication load imposed by the process. The model captures some known aspects of human auditory processing but has not been tested in detail. More importantly, there are a number of important aspects of multi-operator situations that have not been considered. For example: What is one crew member's internal model of another crew member (or of the internal model of that crew member)?; What about social interactions among the crew?; How do team members work jointly to achieve the same goal? These issues are likely to be paramount in unstructured situations but, in the more structured cases which are usually of interest, the crew members have well-defined roles and the problems can be side-stepped for a significant portion of the analysis. Nonetheless, they are issues for research.

We close the discussion with some final comments reflecting the author's perspectives and views in regard to this approach to modelling supervisory control. First, it should be stated that although the control-theoretic approach has been emphasized, it is clear that modelling human supervisory control will be an interdisciplinary venture involving psychologists and computer scientists as well as control theorists and systems specialists. Second, the approach leads to a simulation model for the operator and there will be a great temptation to include everything one can think of with the possibility of winding up with a "fiddler's paradise" in terms of model parameters. This temptation must be resisted as much as possible, but we must also expect some increase in the number of model parameters as we include more and more aspects of behavior in the model. Third, a generic model has been proposed in an effort to provide a somewhat unified framework for modelling human supervisory control behavior. It is recognized, however, that in specific applications, we can expect some aspects of this model to be emphasized and even elaborated while others are deemphasized or ignored. In this process, specific models may bear small resemblance to the generic model. It should be noted that a reduction in model parameters may result from the specialization to a particular application.

Lastly, we repeat that experimental validation tests of the overall supervisory control models (and of some sub-models) have not been conducted. Moreover, the collection of validation data for these models will almost always present major technical and economic problems so that the availability of such data in a timely manner is highly problematical. Because of this, and because of the inherent complexity of the problem, the kind of predictive validity attainable with the OCM is probably too much to hope for in the foreseeable future. In the author's view, continuous manual control models are probably unique in the degree to which they can be validated and, if we are to address the man-machine issues of complex systems in the systematic fashion possible with a quantitative model, we will have to settle for something less. In particular, we should proceed largely from well-reasoned constructs and sub-models and data where we have it in developing supervisory control models. With this level of model validity, we can still make important contributions to system design and regulation by predicting the consequences that obtain *if* the operator behaves in the manner assumed, even though we may be unable to predict actual operator behavior with precision and full confidence.

ACKNOWLEDGEMENT

The general approach to modelling supervisory control described in this paper, as well as specific elements of it, owes much to the following colleagues at BBN: Alper Caglayan, Carl Feehrer, Roy Lancraft, Bill Levison, Ramal Muralidharan, Dick Pew, and Greg Zacharias. They do not, however, share the "responsibility" for the conjectural and philosophical musings of the author.

NOTES

1.   Of course, there are many formulations of specific supervisory tasks or sub-tasks, such as monitoring or decision-making. However, to the author's knowledge, there does not exist a formal statement of supervisory control problems, such as may be given, for example, for wide classes of optimal control problems.

2.   It is interesting to note that in psychology there has been a similar shift to internal model constructs and away from purely stimulus-response (input-output) representations. Clearly, these trends in system theory and psychology reflect the influence of modern computers. In recognizing this similarity, we can see that some of the concerns with "identifiability" of the OCM parallel (Phatak, 1977; Baron, 1977) those of the behaviorists in respect to cognitive modelling.

3.   The central importance of the estimator in the OCM has been recognized for some time. Indeed, the paper that introduced the OCM structure was entitled "The Human as an Optimal Controller and Information Processor" (Baron and Kleinman, 1969).

4.   For discussion purposes, we will describe this process as though it was discrete and sequential. This is likely to be appropriate for supervisory control modelling but it should be noted that in the OCM the process is continuous and simultaneous.

5.   This leads to the intriguing possibility of the estimator becoming so confident in its priors that it virtually ignores the data or, at least, requires significant amounts of "disconfirming" data to change its estimate. This behavior, which actually can be optimal under the right circumstances, is well-known in human observers.

6.   Here, we do not discuss control-theoretic models for visual scanning such as may be found in Baron and Kleinman (1969) or Kleinman and Curry (1977).

7.   This type of failure representation is more general than it might, at first, appear. For example, Caglayan (1980) has shown how actuator, sensor and component failures in a physical system can be modelled by bias jumps in system inputs, outputs and states and how such a failure model results in a bias in the residual sequence. It is also the case that excessive unanticipated disturbances such as wind shears in aircraft operation can be detected on the basis of a non-zero mean residual (Levison, 1978).

8. This is implemented by defining an instantaneous attractiveness measure for each task which balances the expected reward for performing a task against the potential losses for not working on the remaining tasks.

9. The DDM is considered here as modelling the single task, or decision, of what to do next.

10. Two other control theoretic models that address some supervisory control issues, but are not discussed here, are the models of Kok and Stassen (1980) and of Govindaraj and Rouse (1981).

11. It was assumed that the nominal state was known exactly so all uncertainty arose in estimating the perturbation. It is possible to relax this assumption.

12. The probabilities for questions 2 and 3 are related to the reliability that the events actually indicate the situation of interest.

13. Other types of sensory inputs (vestibular, tactile, olfactory) may be considered where appropriate.

14. Of course, estimation performance will generally improve with increasing fidelity of the model.

15. Clearly, this detection probability could be made a parameter.

16. The notion of templates and template matching for diagnosis was suggested to the author by Carl Feehrer.

17. We have in mind postulating new mental model(s) for the system and processing the residuals, in the spirit of the multi-model hypothesis testing algorithms. However, it should be noted that each alternative hypothesis may not require a full n-state mental model (Caglayan, 1980).

18. The decision analysis in Appendix D of that report is due to Ramal Muralidharan.

19. Discrete event simulation models do not model either the cognitive processes or the temporal interactions between human and system in sufficient detail for our purposes.

REFERENCES

Baron, S. A model for human control and monitoring based on modern control theory. *Journal of Cybernetics and Information Sciences 1*(11):3–18, 1976.

Baron, S. Some comments on parameter identification in the optimal control model. *Systems, Man and Cybernetics Review 6*(1):4–6, 1977.

Baron, S., Feehrer, C., Muralidharan, R., Pew, R., and Horwitz, P. An approach to modeling supervisory control of a nuclear power plant. NUREG/CR-2988, ORNL/SUB/81-70523/1, Oak Ridge National Laboratory, Oak Ridge, Tennessee, 1982.

Baron, S., and Kleinman, D. L. The human as an optional controller and information processor. *IEEE Trans. Man-Machine Systems, MMS-10*:9–16, 1969. (Also, NASA CR-1151, Sept. 1968.)

Baron, S., and Levison, W. H. The optimal control model: Status and future directions. *Proceedings of IEEE Conf. on Cybernetics and Society*, Cambridge, MA, 1980, pp. 90–100.

Baron, S., Zacharias, G., Muralidharan, R., and Lancraft, R. PROCRU: A model for analyzing flight crew procedures in approach to landing. *Proceedings of Eight IFAC Work Congress*, Tokyo, Japan, Vol. XV, pp. 71–76, 1980. (Also, NASA Report No. CR152397.)

Caglayan, A. K. Simultaneous failure detection and estimation in linear systems. *Proceedings of the 19th IEEE Conference on Decision and Control*, Albuquerque, N.M., Vol. 2, 1980, pp. 1038–1041.

Chow, E. Y. Failure detection system design methodology. LIDS-TH-1055, MIT, Cambridge, MA, 1976.

Gai, E. G., and Curry, R. E. A model of the human observer in failure detection tasks. *IEEE Trans on Systems, Man and Cybernetics SMC-6:* 85–94, 1976.

Gai, E. G., and Curry, R. E. Failure detection by pilots during automatic landing: Models and experiments. *AIAA J. Aircraft 14* (2), 1977.

Gelb, A., Kasper, J. F., Nash, R. A. Jr., Price, C. F. Jr., and Sutherland, A. A. Jr. *Applied Optimal Estimation.* Cambridge, MA: MIT Press, 1974.

Govindaraj, T., and Rouse, W. B. Modeling the human controller in environments that include continuous and discrete tasks. *IEEE Trans. on Systems, Man and Cybernetics SMC 11* (6):410–417, 1981.

Green, D. H., and Swets, J. A. *Signal Detection Theory and Psychophysics.* New York: John Wiley and Sons, Inc., 1966.

Johannsen, G., and Rouse, W. B. Mathematical concepts for modeling human behavior in complex man-machine systems. *Human Factors 21* (6):733–747, 1979.

Kleinman, D. L., and Curry, R. E. Some new control theoretic models for human display monitoring. *IEEE Trans. on Systems, Man and Cybernetics SMC-7* (11):778–784, 1977.

Kok, J. J., and Stassen, H. G. Human operator control of slowly responding systems: Supervisory control. *Journal of Cybernetics and Information Science 3:*123–174, 1980.

Lane, N., Strieb, M. I., Glenn, F. A., and Wherry, R. J. The human operator simulator: An overview. *Proceedings of Conference on Manned Systems Design*, Freiburg, W. Germany, September, Vol. 1, 1980, pp. 1–39.

Levison, W. H. Analysis and in-simulator evaluation of display and control concepts for a terminal configured vehicle in final approach in a windshear environment. NSAS CR-3034, 1978.

Levison, W. H., Elkind, J. E., and Ward, J. L. Studies of multi-variable control systems: A model for task interference. NASA CR-1746, 1971.

Levison, W. H., and Tanner, R. B. A control theory model for human decision making. NASA CR-1953, 1971.

Luce, R. D., and Galanter, E. Discrimination. In R. D. Luce, R. R. Bush and E. Galanter (Eds.), *Handbook of Mathematical Psychology.*, Vol. 1. New York: John Wiley and Sons, Inc., 1963.

Miller, R. A., and Seifert, D. J. Combined discrete network-continuous control modeling of operator behavior. *Proceedings of Sixteenth Annual Conference on Manual Control*, Cambridge, MA., 1980.

Muralidharan, R., Baron, S., and Feehrer, C. E. Decision, monitoring and control model of the human operator applied to an RPV control problem. BBN Report No. 4075, Cambridge, MA., 1979.

Muralidharan, R., and Baron, S. DEMON: A human operator model for decision making, monitoring and control. *Journal of Cybernetics and Information Science 3:*97–122, 1980.

Pattipati, K. R., Ephrath, A. E., and Kleinman, D. L. Analysis of human decision making in multi-task environments. Tech. Rept. EECS TR-79-15, University of Connecticut, 1979. (Also, *Proceedings of Int. Conf. on Cybernetics and Society*, 1980.)

Pew, R. W., and Baron, S. Perspectives on human performance modelling. *Proceedings of IFAC/IFIP/IFORS/IEA Conference on Analysis, Design and Evaluation of Man-Machine Systems*, (G. Johannsen and J. E. Rynsdorp, Eds.), Baden-Baden, Fed. Republic of Germany, 1982, pp. 1–14.

Pew, R. W., Baron, S., Feehrer, C. E., and Miller, D. C. Critical review and analysis of performance models applicable to man-machine systems evaluation. Report No. 3446, Bolt Beranek and Newman, Inc. Cambridge, MA., 1977.

Phatak, A. V. Formulation and validation of optimal control theoretic models for the human operator. *Systems, Man and Cybernetics Review* 6(1):3,4, 1977.

Phatak, A. and Kleinman, D. L. Current statues of models for the human operator as a controller and decision maker in manned aerospace systems. *Automation in Manned Aerospace Systems*, AGARD Proceedings No. 114, Dayton, Ohio, 1972.

Proceedings of Workshop on Cognitive Modeling of Nuclear Plant Control Room Operators. NUREG-CP-0042 (to appear), Dedham, MA., 1982.

Rasmussen, J. The humans as a systems component. In H. T. Smith and T. A. G. Green (Eds.), *Human Interaction with Computers*. London: Academic Press, 1980a.

Rasmussen, J. Models of mental strategies in process plant diagnosis. *Preprints of Conf. on Human Detection and Diagnosis of System Failures*, Roskilde, Denmark, 1980b.

Rasmussen, J. and Rouse, W. B. (Eds.). *Human Detection and Diagnosis of System Failures*. New York: Plenum Press, 1981.

Rouse, W. B. Models of human problem solving: Detection, diagnosis and compensation for system failures. *Proceedings of the Conference on Analysis, Design and Evaluation of Man-Machine Systems*, Baden-Baden, Germany, 1982.

Sheridan, T. B. Toward a general model of supervisory control. In *Monitoring Behavior and Supervisory Control*. New York and London: Plenum Press, 1976, pp. 271–282.

Sheridan, T. B. and Johannsen, G. *Monitoring Behavior and Supervisory Control*. New York and London: Plenum Press, 1976.

Siegel, A. I., and Wolf, J. J. *Man-Machine Simulation Models*. New York: John Wiley and Sons, Inc., 1969.

Singleton, W. T. The model-supervisor dilemma. *Monitoring Behavior and Supervisory Control*, (T. B. Sheridan and G. Johannsen, Eds.). New York: Plenum Press, 1976, pp. 261–270.

Swain, A. D. and Guttmann, H. E. (Eds.). *Handbook of Human Reliability Analysis with Emphasis on Nuclear Power Plant Applications*. U. S. Nuclear Regulatory Commission, NUREG/DR-1278, 1980.

Tulga, M. K. and Sheridan, T. B. Modeling human decision-making behavior in supervisory control. *Proceedings of Thirteenth Annual Conference on Manual Control.*, MIT, Cambridge, MA, 1977, pp. 199–209.

Wewerinke, P. A model of the human observer and decision maker. *Proceedings of the Seventeenth Annual Conf. on Manual Control.*, Pasadena, CA, 1981.

Willsky, A. S. A survey of design methods for failure detection in dynamic systems. *Automatica* 12, 1976.

Zacharias, G. L., Baron, S., and Muralidharan, R. A supervisory control model of the AAA crew. *Proceedings of 1982 American Conference on Automatic Control.*, Wash. D.C., 1982, pp. 301–311.

SUPERVISORY CONTROL OF REMOTE MANIPULATORS, VEHICLES AND DYNAMIC PROCESSES:

EXPERIMENTS IN COMMAND AND DISPLAY AIDING

Thomas B. Sheridan

ABSTRACT

This chapter is about supervisory control, an increasingly prevalent form of man-machine system wherein a human operator controls a process as the supervisor of a computer. The computer, in turn, may perform limited automatic control or it may process and display information from sensors. The particular context of interest here is supervisory control of manipulators and vehicles for remote inspection and work in the deep ocean.

After giving a more detailed definition of supervisory control and providing examples, the chapter reviews a number of experimental studies conducted recently at the MIT Man-Machine System Laboratory. These

Advances in Man-Machine Systems Research, Volume 1, pages 49–137.
ISBN: 0-89232-404-X

are divided into two groups. The first group of studies is concerned with computer mediation in command and control of manipulation. Two developed systems are described by which an operator may "teach" a manipulator to perform simple manipulation tasks. Other experiments related to the special problems of communicating geometric information, to compensating for motion disturbances, and to the operator's sometime dilemma between allocating an automatic control system vs. doing the task himself.

The next group of experimental studies examines computer mediation in processing sensed information and displaying it to the human supervisor. The first experiment deals with effects on manipulator control of tradeoffs between frame-rate, resolution and grayscale under severe bandwidth constraints. Three subsequent experiments treat the use of computer-generated models to aid planning and real-time control under conditions of limited feedback or time delay. Two final experiments are concerned with aiding the operator in detecting and locating failures.

A brief conclusion reviews how these experiments fit together and speculates on problems and prospects for supervisory control in manipulation, vehicle and process control, and other areas.

I. DEFINITION AND EXAMPLES[1]

A.   What Is Supervisory Control

1.   Definition

Supervisory control of a process means a human operator communicates with a computer to gain information and issue commands, while the computer, through artificial sensors and actuators, implements these commands to control the process. Thus a restrictive use of the term supervisory control means that one or several human operators are setting initial conditions for, intermittently adjusting, and receiving information from a computer which itself closes a control loop through external sensors, effectors and the task environment. We may call this definition A.

However, when the computer makes a sufficiently complex transformation of environmental data to produce integrated (chunked) displays, and/or retransforms operator commands to implement sufficiently detailed control actions—even though there is not intermediate loop closure—one may call this supervisory control also (definition B). The essential difference of (B) from (A) is that in (B) the computer cannot act on new information without new authorization by the supervisor (i.e., the computer implements discrete sets of instruction open loop), though the two cases may look similar to the supervisor who still sees and acts through his computer (analogous to his staff). That is, the supervisor may not know whether his subordinates act open-loop or closed-loop.

Any process can be brought under human supervisory control and thus be subsumed under this definition, including vehicles (aircraft, spacecraft, ships, submarines, ground vehicles of all kinds), continuous product processes (oil, chemical, fossil and nuclear power plants), discrete product processes (manufacturing, construction, farming), robotic/teleoperator devices where not included above, and information processing of all kinds (air traffic, military command and control, office automation, etc.).

Figure 1 characterizes supervisory control in relation to manual control and automatic control. Common to the five man-machine system diagrams are displays and controls interfaced with the human operator, and sensors and actuators interacting with a process or "task." The first two systems on the left represent manual control. (1) is without computer aiding while in (2) significant computer transforming or aiding is done in either or both sensing and acting (controlling) loops. Note that in both (1) and (2) all control decisions depend upon the human operator. When either the minor (3) or major (4) fraction of control is accomplished by control loops closed directly through the computer we call this supervisory control. If, once the control system is set up, essentially all the control is automatic (5), that is, if the human operator can observe but cannot influence the process (other than pulling the plug), it is no longer supervisory control.

The five diagrams are ordered with respect to degree of automation. The progression is not meant to imply either degree of sophistication or degree of desirability.

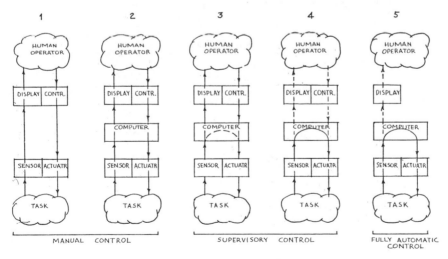

Figure 1. Five Man-Machine Systems Ordered as to Degree of Automation.

2. A Model

Figure 2 shows a more general model of a supervisory control system than Figure 1. The human component is still left as a single entity. There are two subsystems, the human-interactive subsystem (HIS) and the task-interactive subsystem (TIS). The HIS generates requests for information from the TIS and issues high level commands to the TIS (subgoal statements, instructions on how to reach each subgoal or what to do otherwise, and changes in parameters). The TIS, insofar as it has subgoals to reach, instructions on how to try or what to do if it is impeded, functions as an

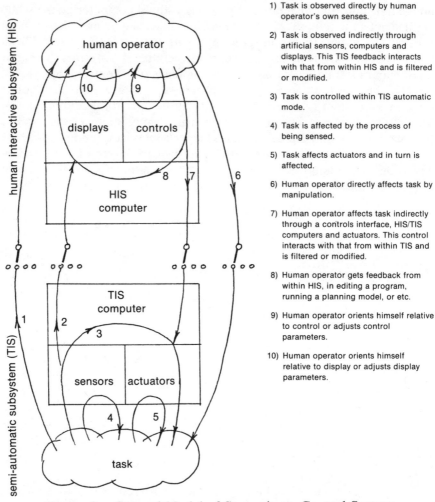

1) Task is observed directly by human operator's own senses.

2) Task is observed indirectly through artificial sensors, computers and displays. This TIS feedback interacts with that from within HIS and is filtered or modified.

3) Task is controlled within TIS automatic mode.

4) Task is affected by the process of being sensed.

5) Task affects actuators and in turn is affected.

6) Human operator directly affects task by manipulation.

7) Human operator affects task indirectly through a controls interface, HIS/TIS computers and actuators. This control interacts with that from within TIS and is filtered or modified.

8) Human operator gets feedback from within HIS, in editing a program, running a planning model, or etc.

9) Human operator orients himself relative to control or adjusts control parameters.

10) Human operator orients himself relative to display or adjusts display parameters.

Figure 2. General Model of Supervisory Control System.

automaton. It uses its own artificial sensors and actuators to close the loop through the environment and do what is commanded.

Note that the HIS and TIS form mirror images of one another. In each case the computer closes a loop through mechanical displacement (hand control, actuator) and electro-optical or sonic (display, sensor) transducers to interact with an external dynamic process (human operator, task). The external process is quite variable in time and space and somewhat unpredictable.

The numbered arrows identify individual cause-effect functions, with explanations of the loops at the right. It is seen that there are three types of inputs into the human operator: (1) those which come by loop 1 directly from the task (direct seeing, hearing or touching), (2) those which come by loops 2 and 8 through the artificial display and are generated by the computer and (3) those which come by loops 10 and 9 from the display or manual controls without going through the computer (i.e. information about the display itself such as brightness or format, present position of manual controls, which is not information which the computer has to provide). Similarly, there are three types of human outputs: (1) those which go by loop 6 directly to the task (the human operator by-passes the manual controls and computer and directly manipulates the task, makes repairs etc.) (2) those which communicate instructions via loops 7 and 8 to the computer, and (3) those which modify the display or manual control parameters via loops 10 and 9 without affecting the computer (i.e., change the location, forces, labels or other properties of the display or manual control devices).

Correspondingly there are three types of force and displacement input into the task: (1) direct manipulations by the operator via loop 6; (2) manipulations controlled by the computer via loops 3 and 7; and (3) those forces which occur by interaction, over loops 4 and 5, with the sensors and actuators and are not mediated or necessarily do what was intended by the computer or operator. Finally there are three types of outputs from the task: (1) information fed back directly to the operator over loop 1; (2) information fed to the TIS computer via loops 2 and 3; and (3) information (in the form of forces and displacements) which modifies the sensors or actuators via loops 4 and 5 without being explicitly sensed by the computer.

When the task is near to the operator, the HIS and TIS computers can be one and the same. When the TIS is remote usually HIS and TIS computers are separated to avoid problems caused by bandwidth or reliability constraints in telecommunication, loops 2 and 7. This problem will be discussed in detail in Section IV.C.

Multiplexing switches are shown in loops 1,2,7, and 6 to suggest that one HIS may be time-shared among many TIS, i.e., many tasks, each with its own local automatic control or robotic implementer. In fact, more and more this is coming to be the case in supervisory control. In some process plants there are over 1000 TIS, some being primitive feedback controllers, some being simply programmed but highly reliable micro-computers. The sheer number of TIS causes a multiplexing or switching overhead cost.

The above is a descriptive model of supervisory control; that is, it is intended to fit what is observed to be the structural and functional nature of a wide variety of situations we discussed earlier. The variables on the lines of Figure 1 are all measurable; there are no intervening variables, no suppositions about what is going on that we cannot observe readily.

Being a descriptive model this is by definition not a normative model. We have not imposed any notions of how the system *should* work, or of within one task and switching tasks. The remainder of this section elaborates this point.

It is important to note, also, that we do not intend to develop a model of the human operator independent of the rest of the system. McRuer and Krendel (1965) abandoned trying to model the human operator in a simple control loop as an invariant entity *per se* and turned instead to finding invariance in the series combination of human controller plus controlled process. It seems that it is best to find invariance in supervisory control plus tool box of computers, sensors and effectors plus task. One may note various functions that neither human *or* computer *can* do, but some are best done by human and some are best done by computer. Which does what function will evolve over many years in the future and will always depend on circumstance. For now the intent is to provide a qualitative description of the combination.

The essence of supervision, as noted in conjunction with the dictionary definition earlier, is that it is not a single activity, as we are accustomed to characterize various sensory-motor or cognitive or decision-making skills, or communication or controlling behavior. Supervision implies that the primary or direct activity, whatever it is, is normally being done by some entity (man or machine) other than the supervisor. There may be a single primary task, or many such tasks. The supervisor from outside, performs those many functions necessary to insure that the single entity does (or multiple entities do) what he, the supervisor, intends. Thus there may be multiplicity of function for two reasons:

1. For each primary task there are many different things to do to ensure that the primary entity (what was called the TIS in the model description) does what the supervisor intends that it do.
2. When there are multiple primary tasks, while the basic functions

may be similar from one primary task to another, the data are different, and the initial conditions are different in performing the same function on each.

Our supervisory control model shows the supervisory computer to multiplex among or alternately connect to, different TIS's or primary tasks. It also shows multiple connections to and from the human operator. It does not make clear that in switching from one TIS to another the initial conditions ("getting one's bearings") are different with each switch. Nor does it make clear that the human supervisor is continually switching functions even while dealing with a single TIS.

But this seems to be the essence of the supervisor: switching functions within one task and switching tasks. The remainder of this section elaborates this point.

Earlier, in conjunction with the dictionary definition, the ideas of "planning," "programming" and "observing" emerged as different components of "supervising."

Missing explicitly from the earlier dictionary definition but implied nevertheless are two additional functions. The first is taking over from the "other" entity, the TIS in our case, seizing direct control when indirect control by supervision fails. The second is to learn from experience.

Summarizing and elaborating to suit our present context, the supervisor, with respect to each task (and each TIS), must perform five distinct functions listed in Table 1. While the explanation of these five functions in Table 1 is not a consensus and some of these steps are manifest to a greater or lesser degree in any actual supervisory control situation, the point is that the necessary sequencing through these differing functions makes the human supervisory controller essentially different from the human in-the-loop, one-continuous function controller and decision-maker.

As implied above the allocation of attention by the human supervisor is both between functions for a given tasks and between tasks. In skilled or overlearned activities a person can engage in many at once (provided the required sensors and effectors are not overtaxed with respect to simple mechanical or signal processing considerations). Thus one can drive a car, talk, scratch his nose and look for a landmark at the same time. But one cannot do multiple simultaneous tasks each of which requires "new thinking" unless the time requirements are such that one can shift attention back and forth. In view of these facts we initially may characterize the attention allocation of the human supervisor as, first, selecting among alternative tasks to be done, and second, selecting his proper function with respect to that task.

Table 1. Functions of the Supervisor

1. *Plan*
a) be aware of what tasks are to be done, what resources are available, what resources (TIS) are committed to what tasks, and what resources are uncommitted
b) decide on overall goal or goals, including objective function or tradeoffs among goals, and including criteria for handling uncertainties
c) decide on strategy or general procedure, including logic of authority (human, HIS computer, TIS computer) in various situations
d) consider known initial conditions and various combinations of probable inputs and possible actions and their consequences in view of system constraints and capabilities
e) determine best action sequence to do what is intended under various situations
f) decide what is to be considered abnormal behavior including automatic recovery from trouble, and what defaults or contingency actions are appropriate.
2. *Teach* (a, b, c and d could also be considered part of planning)
a) estimate what the computers (HIS or TIS) know of the situation
b) decide how to instruct the HIS to instruct the TIS to execute intended and abnormal actions
c) decide how many of intended and abnormal actions TIS should undertake in one frame, i.e. before further instruction
d) try out part or all of that instruction in (operator's) own mental and/or HIS computer model without commitment to transmit to TIS
e) impart instruction (program) to HIS computer with commitment to transmit to TIS
f) give command to HIS to start action

B. Examples Of Supervisory Control In Remote Manipulation

Intrinsic to supervisory control is the idea of teleoperation—man performs a sensing and/or manipulation task remotely by use of artificial sensors and actuators. This can be spatial remoteness, as with a remotely controlled vehicle or manipulator undersea or in space. It can be temporal remoteness, due to a time delay between when an operator issues com-

3. *Monitor*
   a) decide on what TIS behavior to observe
   b) specify to HIS computer the desired display format
   c) observe display, looking for signals of abnormal behavior and performing on-line computer-aided analysis of trends or prediction or cross-correlation as required
   d) observe task directly when and if necessary
   e) make minor adjustments of system parameters when necessary, as the automatic control continues.
   f) diagnose apparant abnormalities or failure, if they occur, using computer aids

4. *Intervene*
   a) decide when continuation of automatic control would cease to be satisfactory and minor parameter adjustments would not suffice either
   b) go physically to TIS or bypass all or portions of HIS and TIS computers to effect alternative control actions or stoppage or recovery
   c) implement maintenance or repair or modifications of TIS or task
   d) recycle to (1), (2) or (3) as appropriate

5. *Learn*
   a) decide means for collecting salient data and drawing inferences from it over repeated system runs
   b) implement these means
   c) allow for serendipitous learning
   d) periodically take stock of learning, modify system hardware and software, and anticipate future planning of operations
   e) develop understanding of and trust in the system

mands and when he receives feedback. Or it can be functional remoteness, meaning that what the operator sees and does and what the system senses and does bear little superficial resemblance. Teleoperation can be either the motivation for or the result of supervisory control, as will be made evident.

In a sense, manipulators combine the functions of process control and vehicle control. The manipulator base may be carried on a spacecraft, a ground vehicle, or a submarine, or its base may be carried on a spacecraft, a ground vehicle, or a submarine, or its base may be fixed. The hand (gripper, end effector) is moved relative to the base in up to three degrees of translation and three degrees of rotation. It may have one degree of freedom for gripping, but some hands have differentially movable fingers or otherwise have more degrees of freedom to perform special cutting, drilling, finishing, cleaning, welding, paint spraying, sensing, or other functions.

Manipulators are being used in many different applications, including lunar roving vehicles, undersea operations, and hazardous operations in industry. The type of supervisory control and its justification differ according to the application.

The fact of a three-second time delay in the earth-lunar control loop resulting from round-trip radio transmission from earth leads to instabilities, unless an operator waits three seconds after each of a series of incremental movements. This makes direct manual control time-consuming and impractical. Sheridan and Ferrell (1967) proposed having a computer on the moon receive commands to complete segments of a movement task locally using local sensors and local computer program control. They proposed calling this mode supervisory control. Delays in sending the task segments from earth to moon would be unimportant, so long as rapid local control could introduce actions to deal with obstacles or perform other self-protection functions.

The importance of supervisory control to the undersea vehicle manipulator is also compelling. There are things the operator cannot sense or can sense only with great difficulty and time delay (e.g., the mud may easily be stirred up, producing turbid opaque water that prevents the video camera from seeing), so that local sensing and automatic response may be more reliable. For monotonous tasks (e.g., inspecting pipelines, structures, or ship hulls or surveying the ocean bottom to find some object) the operator cannot remain alert for long; if adequate artificial sensing could be provided for the key variables, supervisory control should be much more reliable. The human operator may have other things to do, so that supervisory control would facilitate periodic checks to update the computer program or help the remote device get out of trouble. A final reason for supervisory control, and often the most acceptable, is that, if communications, power, or other systems fail, there are fail-safe control modes into which the remote system reverts to get the vehicle back to the surface or otherwise render it recoverable.

Many of these same reasons for supervisory control apply to other uses of manipulators. Probably the greatest current interest in manipulators is for manufacturing (so-called industrial robots), including machining, welding, paint spraying, heat treatment, surface cleaning, bin picking, parts feeding for punch presses, handling between transfer lines, assembly, inspection, loading and unloading finished units, and warehousing. Today repetitive tasks such as welding and paint spraying can be programmed by the supervisor, then implemented with the control loops that report position and velocity. If the parts conveyor is sufficiently reliable, welding or painting nonexistent objects seldom occurs, so that more sophisticated feedback, involving touch or vision, is usually not required. Manufacturing assembly, however, has proven to be a far more difficult task.

In contrast to assembly line operations, in which, even if there is a mix of products, every task is prespecified, in many new applications of manipulators with supervisory control each new task is unpredictable to a considerable extent. Some examples are mining, earth moving, building construction, building and street cleaning and maintenance, trash collection, logging, and crop harvesting, in which large forces and power must be applied to external objects. The human operator is necessary to program or otherwise guide the manipulator in some degrees of freedom, to accomodate each new situation; in other respects certain characteristic motions are preprogrammed and need only to be initiated at the correct time. In some medical applications, such as microsurgery, the goal is to minify rather than enlarge motions and forces, to extend the surgeon's hand tools through tiny body cavities to cut, to obtain tissue samples, to remove unhealthy tissue, or to stitch. Again, the surgeon controls some degrees of freedom (e.g., of an opitcal probe or a cauterizing snare), while automation controls other variables (e.g., air or water pressure).

II. ILLUSTRATIVE EXPERIMENTS ON COMMAND AIDS TO THE SUPERVISOR

The experiments described below are concerned with the "effector" loops of Figure 2 (loops 6 and 7) and the automation (loop 3) which augments operator control. Sections II.A and II.B describe the development of particular supervisory command/control systems for remote manipulation. Section II.C examines the problems of "pointing," that is, telling the computer about the specific geometrical positions and orientations in the world which it must know in order to do its job; that such nominally motor activity will be shown to be highly influenced by perceptual factors points up the artificiality of trying to separate effector from sensory mechanisms. Section II.D exemplifies a simple form of automatic error nulling within a manual loop, a primitive form of supervisory control. Section II.E discusses some relatively abstract laboratory research dealing with a new dilemma facing the supervisor—when should he allocate a machine to do a task and when should he do it himself.

A. Supervisory Command Of Remote Manipulation:
"SUPERMAN"[2]

1. Introduction

To investigate the relative merits of supervisory control applied to teleoperators, specifically telemanipulators in this case, a task-referenced
sensor-aided supervisory system, called SUPERMAN, was built and experiments were performed. These experiments compare various conventional control modes with supervisory control, and demonstrate that supervisory manipulation does improve performance in the majority of
cases.

2. Method and Apparatus

The major elements of the SUPERMAN system are a modified Argonne
E2 master-slave manipulator with six degrees-of-freedom, a dedicated
control interface (DASI), and an Interdata 70 computer. Designed for
efficient man-machine interaction with both analog and symbolic control
inputs, the system can be commanded by a variety of conventional control
modes as well as supervisory. In addition, time delay and/or noise can
be added for experimental purposes.

Using both analog and symbolic commands, a manipulation can be
taught and/or demonstrated to the computer. Trained manipulations can
be transformed from one coordinate system to another so that once the
generic characteristics of a task have been learned, the machine can perform similar tasks in different locations without further training. When
the human operator requires a particular trained manipulation he simply
"initializes" the new coordinate system relative to the old by moving the
teleoperator hand to the starting point of the task (e.g., grasping a nut or
valve handle) and signals for execution. Certain objects in the task environment can, of course, maintain their original coordinates.

Since the E-2 manipulator can sense the forces generated during the
task, supervisory programs can call for repeated movements which, upon
certain touch conditions becoming true, branch into other movements.
For example, repeated hand movements can grasp a nut, unscrew it by
one revolution, pull back to test whether it is off and, if it is, place it in
a bucket or, if it is not, repeat the operation. Similar supervisory programs
have been applied to attaching a nut to a bolt, opening and closing a valve,
scooping dirt and so on. Further information on the SUPERMAN system
can be found in Brooks (1980).

The manipulator laboratory was arranged as shown in Figure 3a during
the experiments. To simulate remote conditions the operator viewed the

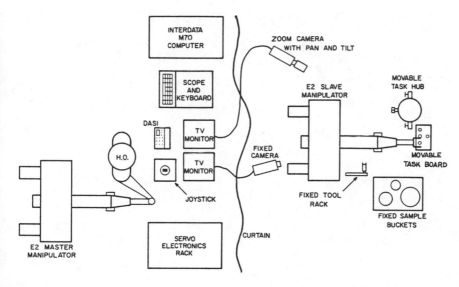

Figure 3a. Schematic of Experimental Layout.

task environment through either a mono or 2-view television system. The video system consisted of two black and white high-resolution 9 in. monitors, a fixed camera with wide angle lens, and a zoom camera with pan and tilt.

Figure 3b shows the manipulator environment and the experimental tasks designed for this study. The tool rack and sample buckets remained in the locations shown throughout the experiments since these pieces of equipment are usually rigidly attached to the teleoperator vehicle in real applications. Also shown in the figure are the movable task hub and task board on which representative tasks such as valves, bolts, etc. were mounted. The location of the task hub and board were changed throughout the study to simulate the random task/vehicle relationships which are typical of the arbitrary environments found in marine and space applications.

3. Experimental Design

Six manipulation tasks were identified for Experimental Investigation: (1) tool retrieval; (2) tool return; (3) taking a nut off; (4) grasping an object and placing it in a container; (5) opening/closing a valve; and (6) digging. In addition, four manual control modes were delineated as important experimental parameters: (1) switch fixed rate; (2) joystick variable rate; (3) master-slave position control; and (4) master-slave position control with force reflection. With regard to the video arrangement, both mono and

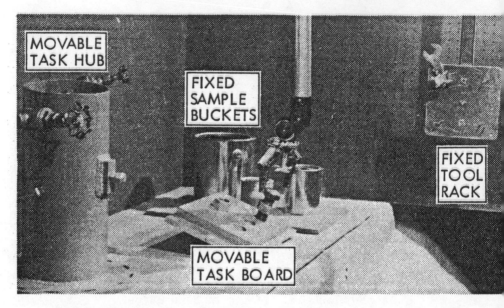

Figure 3b. Task Hub, Task Board, Sample Buckets, and Tool Rack.

2-view conditions were tested for comparison. Due to time constraints only three subjects were used for four of the tasks (tool retrieval, tool return, nut-off and sampler), and only one subject was used for the remaining two (open/close valve and digger). Each experiment was performed 5 times by each subject to obtain a statistical mean and standard deviation. Both manual and supervisory control were used.

These conditions result in a total of 1120 experimental runs. Since this would require an inordinate amount of time, the experimental load was reduced to 680 runs by noting that some of the tasks, or portions of the tasks, had constant computer execution times.

4. Subjects and Training

Three classes of subjects were used for these experiments, one experienced, four well trained, and two untrained subjects. The experienced subject had over 200 hours of training of this particular system. The well trained subjects had an average of 20 hours training given in 15 minute intervals for each of the control modes.

5. Procedure

The experiments were scored on the basis of recorded time and errors. The subjects were not given specific instructions to minimize either quality, but only to weigh them equally.

Tool-Retrieval Task—The first task required the subject to start with the end effector positioned near the task hub. On the experimenter's signal, the subject moved the end effector to the tool rack, obtained the tool, being sure it was properly seated in the hand, and returned with the tool to the starting position. The subjects were told that the success or failure of the task was measured by whether a solid connection between the tool handle and end effector was achieved. Execution of this task under supervisory control simply involved a button push.

Tool-Return Task—For the second task the subject started from a position next to the task hub with the tool in hand, and on the experimenter's signal, moved to the rack, replaced the tool insuring that it was properly seated, and returned to the initial position. The operators were told that the success or failure of the task was determined by whether or not the tool was properly replaced on the rack. To properly seat the tool on the rack required that both of the 1/8 inch rack pins were engaged in the handle and that the tool was completely pushed onto the pins. This task was executed under supervisory control through a simple button push.

Nut-Removal Task—This experiment began with the end effector positioned over the valve on the task hub. On the experimenter's signal, the subject moved the end effector from the valve to the nut, oriented the hand, and removed the nut. The general procedure used by the subjects and computer was to turn 180°, pull back to test if the nut was off, and then either reverse 180° and continue, or remove the nut. Prior to the task, the operators were told that the task would be considered successfully completed if the nut could be removed without losing it. Under supervisory control the operator initialized the task by moving from the starting position to the nut, orienting the hand with the rotational axis of the nut, and signaling the computer to remove it.

Sampling Task—The fourth task required the subject to pick-up thirteen randomly placed samples and put them in one of two buckets according to their size. The subjects were told that their success or failure to complete the task would be measured by how many samples were successfully placed in the proper buckets. Under supervisory control the operator initialized the task by placing the end effector over the sample and signaling the computer to place it in the appropriate bucket. The computer returned control to the subject at the location where the sample was grasped. The operator then moved to another sample, initialized, and continued until all 13 samples were in the buckets.

Open/Close Valve Test—This experiment required the subject to position the end effector over the nut on the task hub, and then, on the experimenter's signal, the subject moved to the valve, oriented the hand, and opened or closed the valve as required (opening and closing tasks were switched after each experiment). The subject was required to continue until the valve operation was complete. To initialize this task under supervisory control the operator oriented the end effector on the rotational axis of the valve and signaled the computer either to open or close it as required. The computer checked the rotational torques to determine if the task had been completed.

Digging Task—The final task required the subject to remove a specified amount of soil from a box by filling a bucket with a shovel. This task is composed of a number of subtasks: (1) the shovel is positioned to remove the soil, (2) the shovel is pushed into the soil and lifted out, and (3) the soil is transported to the bucket and dropped in. The subject was required to continue until the bucket was filled. Under supervisory control the positioning of the shovel was performed manually (i.e., the operator decided when and where to dig) while the scooping and dropping actions were executed by the computer.

6. Results

It has been shown by a number of investigators that the time required to perform a task can be attributed to a number of distinctly different motions. For example, one classification divides the task time for control with a time delay into segments related to get, transport, and position motions. For a peg-in-the-hole task Hill (1976) has shown that there are two independent motions which determine the total task time under manual control—gross travel and precision. This chapter will use a similar scheme to describe the task completion time for a supervisory system:

$$t_{TT} = t_I + t_P$$

where

t_{TT} = Task Time
t_I = Time required by the human operator to *initialize* the task. This time is primarily a function of the initial hand/task locations and the manual control mode used to locate the task.
t_P = Time required by the computer to *perform* the task. This time is primarily a function of the task complexity.

The determination of these times is rather simple due to the discontinuity in control which occurs during the trade from manual initialization to computer execution (this "discontinuity" is a desired result since trading of control should be "apparent").

Figures 4–7 are plots of typical data. The data recorded during the supervisory experiments have been divided into initialization and per-

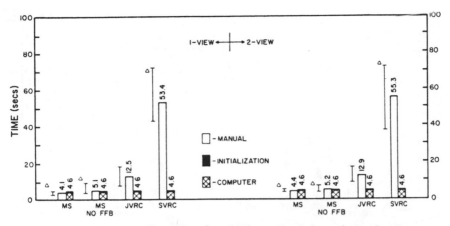

Figure 4a. Average Tool-Retrieval Time. Each bar gives the average time of two subjects. The △ symbol represents the mean time for an untrained subject. The capped lines show the total range of data for the trained subjects.

formance times to indicate the time spent by each action. Each of the time bars is the result of data averaged over two trained subjects, except for Figure 7 which is averaged over three trained subjects. The lines to the left of the manual control bars give the range over which the trained subjects performed the task. For comparison, the average time for an inexperienced subject to perform the first three tasks is also given (denoted by triangles). The mean times of the untrained subjects were always above the maximum value of the trained subjects. The lines to the left of the manual control bars give the range over which the trained subjects performed the task. For comparison, the average time for an inexperi-

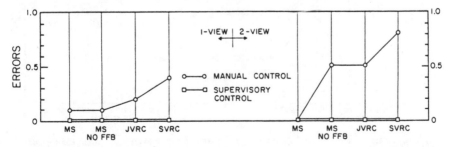

Figure 4b. Expected Number of Tool-Retrieval Errors. Each data point represents the average error rate of two trained subjects. Possible errors included collisions, dropping the tool, and not seating the handle in the end effector properly.

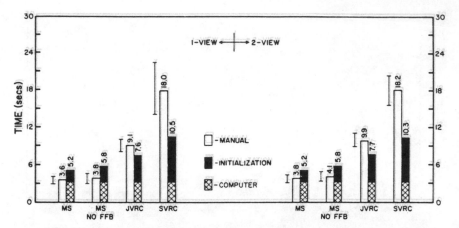

Figure 5a. Average Sampling Time. Each bar represents the mean time of three trained subjects. The capped lines represent the total range of data for the subjects.

enced subject to perform the first three tasks is also given (denoted by triangles). The mean times of the untrained subjects were always above the maximum value of the trained subjects for the same task and control mode. The lower portion of each Figure (Figure 4b–7b) plots the mean number of errors which occurred under manual and supervisory control.

7. Manual Control

Predictably, the task completion time increased with control complexity for all tasks. Viewing conditions (mono and 2-view) appeared to affect

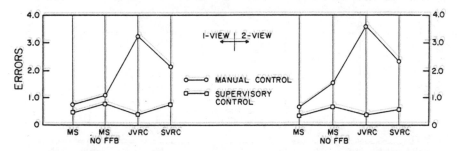

Figure 5b. Expected Number of Sampling Errors. Each data point represents the average error rate of three trained subjects for 13 sampling actions. Possible errors included collisions, missed buckets, lost samples, and (under supervisory control) pressing the wrong button.

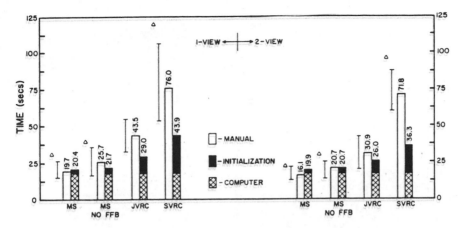

Figure 6a. Average Nut-Removal Time. Each bar represents the
average time of two trained subjects and each △ gives the mean time
for an untrained subject. The capped lines represent the total range of
data for the trained subject.

tasks which required precision movements (e.g., return tool and nut-off),
but had little or no effect on the less precise tasks (e.g., sampling). In
general, the number of errors increased as the control complexity increased from master-slave to switch rate. However, for some of the tasks
a sharp decrease in errors was noticed between joystick and switch rate
control (e.g., see Figure 6b and 7b). This effect is attributable to two
factors: (1) the increased attention and care each operator exhibited during
switch rate control modes (i.e., to move from point A to point B requires
considerable thought and effort with switch rate control, but under joystick control the desired movement only requires a push on the stick), and

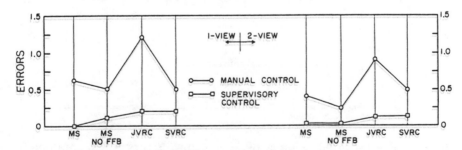

Figure 6b. Expected Number of Nut-Removal Errors. Each data
point represents the average error rate of two trained subjects. Possible
errors included collisions and dropping the nut.

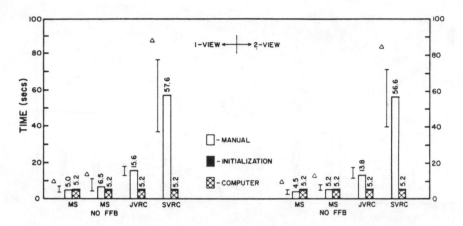

Figure 7a. Average Tool-Return Time. Each bar represents the average time of two trained subjects and each △ gives the mean time for an untrained subject. The capped lines represent the total range of data for the trained subjects.

(2) the coincidental matching of the task degrees of freedom and control degrees of freedom (e.g., in the valve or nut-off tasks the axis of rotation corresponded with the hand axis of rotation).

Table 2 gives the ratio of task completion times for each control mode with respect to the "best" control case, master-slave with force feedback. The ratios are given for each subject, task and viewing condition. The untrained subjects are denoted by U1 and U2, the trained subjects are denoted by T1, T2, T3 and T4, and the experienced subject is denoted by E1. The table shows a number of interesting trends: (1) the ratios

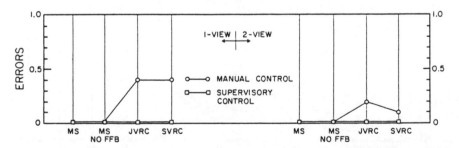

Figure 7b. Expected Number of Tool-Return Errors. Each data point represents the average error rate of two trained subjects. Possible errors included collisions, dropping the tool, and not seating the handle on the rack properly.

Table 2. Ratio of Time to Perform Task Under Given Control Mode to Time to Perform (Task Under Master-Slave with Force Feedback (CM/MS).

	1-VIEW				2-VIEW				
	MS	MS NO FFB	JVRC	SVRC	MS	MS NO FFB	JVRC	SVRC	
VALVE 1-DOF	1.0	1.2	1.4	3.6	1.0	1.1	1.6	3.9	E1
NUT-OFF 2-DOF	1.0	1.2	2.2	4.3	1.0	1.3	2.0	4.4	T1
	1.0	1.4	2.2	3.3	1.0	1.2	1.9	4.5	E1
	1.0	1.3	*	3.9	1.0	1.4	*	4.3	U2
SAMPLER 3-DOF	1.0	1.1	2.5	5.1	1.0	1.1	2.7	4.4	T2
	1.0	1.0	2.8	5.3	1.0	1.0	2.5	5.1	T3
	1.0	1.0	2.4	4.7	1.0	1.1	2.6	4.9	E1
SCOOPER ?-DOF	1.0	1.1	3.9	10.6	1.0	1.2	3.9	10.9	E1
RETURN-TOOL 6-DOF	1.0	1.1	3.6	10.7	1.0	1.2	3.4	12.1	T4
	1.0	1.4	2.7	12.2	1.0	1.3	2.7	13.2	E1
	1.0	1.4	*	9.0	1.0	1.4	*	9.7	U1
GET-TOOL 6-DOF	1.0	1.1	3.8	13.0	1.0	1.2	3.3	12.1	T4
	1.0	1.4	2.4	13.0	1.0	1.2	2.6	13.2	E1
	1.0	1.4	*	8.3	1.0	1.3	*	11.8	U1

increase with increasing control complexity, (2) the ratios are approximately constant across subjects (both trained and untrained) within a given task, (3) the ratios are constant across viewing conditions, and (4) the ratios are not constant across tasks (the tasks have been arranged in the table so that the ratio increases as the page is read from top to bottom).

8. Supervisory Control

As would be expected, the time required by the computer to perform its portion of the task remained fixed regardless of the manual control mode from which the human operator issued the execution command. Also, since the only action required of the operator to initiate the tool-retrieval and return tasks was a button push, the absence of initialization times in Figure 4a and 5a was not surprising. The remaining tasks, including those not shown in this chapter had initialization times associated with the overall task time. As seen in Figure 6 and 7 the initialization times increased with control complexity.

Table 3 gives the ratios of the task completion times under manual control to the times under supervisory control. The ratios are given for

Table 3.   Ratio of Time to Perform Task Under Manual Control to
Time to Perform Task Under Supervisory Control (MC/SC).

	1-VIEW				2-VIEW				
	MS	MS NO FFB	JVRC	SVRC	MS	MS NO FFB	JVRC	SVRC	
VALVE 1-DOF	1.1	1.3	1.2	2.0	0.8	0.9	1.1	2.0	E1
NUT-OFF 2-DOF	1.0	1.2	1.6	2.0	0.8	1.1	1.2	2.1	T1
	0.9	1.2	1.4	1.5	0.8	0.9	1.1	1.9	E1
	1.5	1.9	*	2.1	1.1	1.5	*	2.0	U2
SAMPLER 3-DOF	0.7	0.7	1.2	1.8	0.8	0.8	1.3	1.9	T2
	0.7	0.7	1.2	1.6	0.7	0.7	1.3	1.7	T3
	0.6	0.6	1.2	1.7	0.7	0.7	1.3	1.8	E1
SCOOPER ?-DOF	0.5	.06	1.9	4.8	0.5	0.6	1.9	4.9	E1
RETURN-TOOL 6-DOF	0.9	1.0	3.2	9.4	0.9	1.0	3.0	10.6	T4
	1.1	1.5	2.8	12.8	0.9	1.1	2.3	11.3	E1
	1.9	2.6	*	16.9	1.7	2.4	*	16.3	U1
GET-TOOL 6-DOF	0.8	0.9	3.2	10.8	1.0	1.2	3.2	11.7	T4
	1.0	1.3	2.3	12.5	0.9	1.1	2.4	12.4	E1
	1.8	2.6	*	15.3	1.4	1.7	*	16.0	U1

each subject, task and viewing condition. The ratios relative to computer
control (Table 3) do not show the same trends as those relative to master-
slave control (Table 2). It is interesting to note that in contrast to the
consistent ratios of Table 2, the computer control ratios of the untrained
subjects are significantly higher than the trained subjects: clearly, untrained subjects gain more from supervisory control than trained subjects.
Gains from supervisory control for any manual mode are seen to be most
significant for tasks which do not require initialization procedures other
than a button push (i.e., tool-retrieval and tool-return). The control mode
columns clearly indicate the results of the SUPERMAN experiments: (1)
master-slave with force feedback rarely benefits from supervisory control,
(2) master-slave without force feedback can profit from supervisory control in tasks which require force feedback, and (3) both forms of rate
control can be aided by supervisory routines regardless of the task.

In all cases the error rates for supervisory control were less than manual
control. However, an interesting error was noted during the sampling
experiments—occasionally the subjects pressed an incorrect button sending the sample to the wrong bucket.

Theoretically there is no reason why master-slave with force feedback should be any faster than supervisory control. Consider that the computer could simply mimic the human operator's best trajectory, and hence, be at least as fast. Unfortunately, in practice there is always a certain overhead associated with retransformation of coordinates, trajectory calculations and sensor logic. Also, it was generally observed that the subjects were making adaptive, orchestrated motions, whereas the computer was limited to more rigidly defined trajectories and states. In light of these observations it can be said that the faster master-slave times make more of a statement about the direction that future studies dealing with supervisory control should take than they do about its potential in teleoperator systems.

Although the experiments were not designed to measure the effectiveness of supervisory control during extended periods of manipulation, an interesting observation was made after the experiments had been completed—the manual experiments had been performed with rest periods between each run because the subjects complained of fatigue and boredom, while the supervisory experiments had been unintentionally run back-to-back since fatigue and boredom were not noted. From these observations it could be surmised that as a task becomes more involved and complex, boredom and fatigue will become increasingly important factors, tipping the scales even further in favor of supervisory control. However, experiments to validate this statement have yet to be performed.

9. Conclusions from SUPERMAN Experiments

Even under "ideal" control conditions supervisory control was found to be more efficient and effective (as determined from the task completion times and manipulation errors) than switch rate control, joystick rate control, and master-slave position control. Bilateral force-reflecting master-slave was found to be slightly faster than supervisory control, but more prone to errors. Since the experiments were performed under "ideal" conditions, it can be reasonably predicted that supervisory control will show even more advantage when used with degraded sensor or control loops (e.g., time delays, limited bandwidth, etc.), though the latter experiments remain to be done. In addition, an a posteriori observation of the experimental procedure appears to indicate that the effects of operator fatigue and boredom during extended periods of manipulation can be significantly reduced through supervisory control.

B. An Improved Supervisory Command System: MMIT[3]

1. Introduction

Following Brooks' work Yoerger (1982) built a refined system for supervisory control of manipulators for underwater vehicles, primarily in the context of close-up inspection. The successor system we call MMIT for Man-Machine Interface for Telemanipulation. It consists of a movement control language and a computer graphic interface.

Using the MMIT system, an operator can teach the computer how to execute tasks for which the operator then monitors the system's progress. One of the most important features of the interface is that it can be used by operators with no physical understanding of manipulator control. The MMIT system was not designed to act independently of a human operator or as an automaton, rather it is a method for extending manual control.

This section provides a description of the interface design. It also describes an experiment performed to test MMIT in a simulated remote inspection task. The results from that experiment show great promise for practical applications of the system. Three general results are particularly important:

1. Task performance in terms of accuracy was improved under MMIT control.
2. The interface decreased the operator's dependence on visual feedback.
3. The system decreased the variability in performance between operators.

The MMIT system is currently in use at MIT and Naval Ocean Systems Center, San Diego. It will be used on RECON-V, a submersible being tested at MIT, and is currently a candidate for several other remotely operated vehicles.

2. System Design

There are several important features in the MMIT design. These include an advanced programming system that features analogic teaching, the combining of analogic data with symbolic commands, and the use of a dynamic simulation for visual feedback to the operator.

The programming system used in the interface allows the operator to teach the computer how to perform remote tasks. Tasks may be either preprogrammed, or new programs may be created during actual remote operation. The computer graphic display can be used to help the operator understand elements of the programming system without requiring a formal mathematical description of how the commands work.

The programming system is based on cartesian coordinate frames, as described for teleoperator tasks by Brooks. Absolute coordinate frames (defined relative to the manipulator base) may be entered analogically, by pointing out their location with the manipulator. Relative coordinate frames (to be applied to an arbitrary coordinate frame) may be defined either analogically or symbolically by entering a semantic description.

The analogic teaching capability is important in the system because it allows the operator to establish relevant positions and orientations quickly, transparently, and with minimal sensing on the part of the remote system. The operator moves the manipulator to the desired position and specifies a name for the position. A cartesian coordinate frame is then computed based on the current angles of the manipulator joints and is stored under the give name.

The arm may be repositioned at any of the defined coordinate frames (called a POSITION), or made to traverse along a series of such frames (called a PATH). This capability is useful for defining positions or paths that the arm should return to several times, such as the grasping position of a tool.

Analogic teaching can be especially useful when an operator does not know the exact coordinate values of a position, but can see via the graphic or television display what he wants. Using analogic definitions in combination with symbolic commands simplifies the teaching of teleoperation tasks for the operator. Programs can be written in which the operator points out an object in the task environment, and then describes an operation on the object by specifying motions built on defined positions.

Using the relative commands, the arm may be moved relative to its current position and orientation. These commands are useful for describing tool motions, such as turning a valve or brushing a weld.

Relative motions defined symbolically and absolute coordinate frames defined analogically may be used together very effectively. Using the analogic capabilities, the operator may define an object of interest as a series of coordinate frames. These absolute coordinate frames can then be used as reference frames for relative motions which define a task. An example, inspection of a weld, will be described later.

The computer can be instructed to make decisions during the control process using structured flow of control statements. Complicated tasks for the manipulator can be composed as a hierarchy of subtasks (Figure 8) through the extensibility mechanism of the FORTH language in which the system is implemented. Once the motions required for a particular task have been defined in the program and analogic data has been input regarding the environment, entire tasks can be accomplished with a single command.

A key element in the effective operation of the system is use of the dynamic computer graphic display, designed by Winey and developed further by Fyler (see Sections III.B and III.D, respectively). This display allows the operator to view an image of the manipulator and task environment from any angle. The operator can rotate, translate, or zoom the

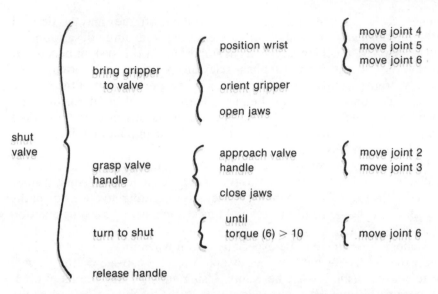

Figure 8.   Warnier-Orr Diagram of Shut Valve task. This is similar to
the procedural net model representation, but decisions may also
included. In this example, the valve will continue to be turned until a
present value of torque is reached on the last manipulator joint.

display. The dynamic display serves three major functions within this
system.

1.   The display is a very effective aid for monitoring the actions of the
manipulator and the condition of its environment. Because the images
are simulated, the quality of the display is not dependent on the quality
of a television picture.

2.   Motions of the manipulator can be simulated to test programs before they are actually executed. Such simulations can be run in faster
than real time.

3.   The display can be used to show the results of computations done
by the computer after high level descriptions have been given, thus helping
the operator understand how the system works. Defined coordinate
frames corresponding to key parts of the environment may also be displayed.

It is important to note that the operator remains in control of the system
during the execution of all tasks. If he detects a problem of which the
system is unaware, he can interrupt and assume control manually or he
can invoke computerized functions to get out of trouble, such as repeating
the last several computer-controlled moves in reverse order.

3. Experimental Evaluation of MMIT

Experimental Design—An experiment was designed to evaluate a working model of the MMIT system. The experiment explored how well an underwater inspection task could be preformed with a master-slave manipulator under three different methods of control. Performance was compared for one manual control mode and two supervisory configurations.

The experimental task was the remote inspection of an underwater weld. This task was chosen because it has proven difficult to perform under manual control in actual underwater tests and includes many representative elements of supervisory control which can be measured and perhaps generalized. Remote weld inspection usually involves two steps. The weld is first cleaned down to bright metal and then some form of non-destructive testing (NDT) is performed, the most common being still photography. Both steps share the same basic requirements—that a tool (cleaning jet or camera) be moved through a trajectory that traverses the weld, that keeps the tool pointed at the weld, and that maintains a specified distance between the tool and the weld.

Performance of the task was evaluated qualitatively in terms of completion time and the error of the manipulator arm in following the desired trajectory.

Three control modes were evaluated in the experiment. The first was standard master-slave with force feedback. The second and third modes supplemented the master-slave control with different types of supervisory computational aids.

Mode 1: *Master-slave control.* This mode is generally considered to be the mode with the best performance of any purely manual control mode. The computer was not used for control.

Mode 2: *Analogic teaching.* In this mode, the operator teaches the trajectory to be followed as a series of discrete positions using the master-slave control. The system can then move the arm through a smooth path which intersects all taught positions. The system does not transform the taught positions, but only interpolates between them, so the computational requirements are quite low. A major advantage over mode 1 is that the taught path may be repeated as desired. This mode is similar to the method used to program many industrial robots.

Mode 3: *Combined analogic and symbolic teaching.* In this mode, the operator first teaches the weld (rather than the trajectory) as a series of discrete positions. The operator then invokes a procedure that symbolically describes a trajectory relative to the taught positions. The procedure attempts to create a trajectory that points at the weld while maintaining a specified distance between the manipulator and the weld. An advantage of this mode over mode 2 is that the trajectory may be repeated as desired with different distances or orientations between the analogic description of the weld and the generated trajectory.

During actual water jet cleaning underwater, the operator has difficulty seeing because the jetting action obscures his view. For this type of task, either of the supervisory modes (2 or 3) would have a distinct advantage over any manual control method by separating the task into teaching and execution phases. For both modes 2 and 3, the operator could do the teaching before the jet obscures his view, then the computer could carry out the jetting operation.

Experimental Setup—The experimental setup is shown in Figure 9. The manipulator used in the experiment was an E-2 master-slave arm. The computer was a PDP 11/34 running the MMIT software. The sensors and actuators of the manipulator were the same for all control modes.

A test weld was developed for the inspection task which included both straight and smoothly curved sections. Subjects viewed the weld and remote manipulator through a television system. The weld and camera were arranged to present a variety of viewing angles to the operator in order to test the sensitivity of the different modes to changing spatial relationships between the camera, manipulator, and weld. The specific task consisted of defining a trajectory which remained one inch from the weld and pointed directly at the weld.

Experimental Trials—Three experienced subjects were used in the experiment. Sessions for each subject consisted of three trials in each control mode with the test weld in three different orientations for a total of nine trials per session. Each subject performed three sessions, and data was recorded on the last session.

Performance Criteria—Accuracy was evaluated by two criteria as shown in Figure 10:

1. The shortest distance between the tip of the manipulator tool and the weld was computed as a function of time, with RMS computed for each run. For perfect performance, this distance would have been maintained at one inch.
2. As a measure of orientation accuracy, the distance between a line oriented with the hand and the closest point on the weld was computed. This distance corresponds to the distance the center of water jet would miss the center of the weld. RMS was computed for each run.

For these measures, estimates of the measurement noise were obtained. Scores for all modes were substantially higher than the measurement noise.

Experimental Results—Analysis of variance showed the effect of control mode was significant for both the position and orientation criteria

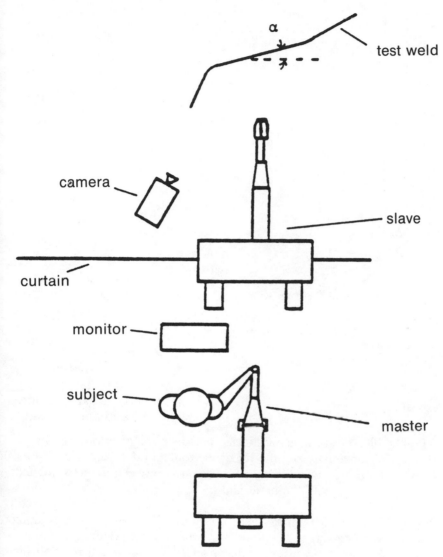

Figure 9. Experimental Setup. Three orientations of the test weld were used (α = $-15°, 0°$, and $15°$).

(p<0.025). Examination of plots of subject means and RMS errors for both position and orientation criteria (Figures 11 and 12) show that each subject improved similarly across the control modes. In general, the largest improvement was between modes 2 and 3. Mode 3 also showed the lowest variation between subjects.

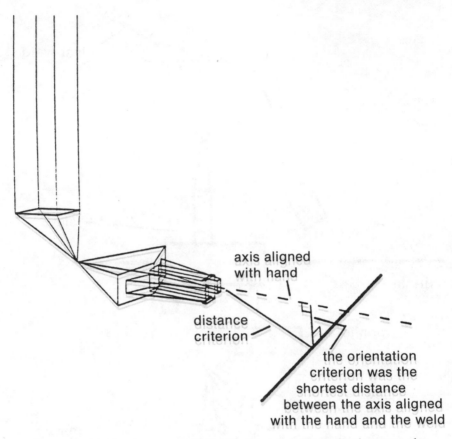

Figure 10. Performance Criteria. The distance criterion was the shortest distance between the tip of the tool and the weld bead. The orientation criterion was the shortest distance between the axis aligned with the hand and the weld bead.

Figure 13 shows the performance criteria as a function of distance along the weld, averaged across all subjects and runs. Large systematic changes in the errors can be seen for modes 1 and 2 as the spatial relationship between the camera and the weld changes. For mode 3, performance was much more uniform despite the large changes in the quality of the visual feedback.

4. Conclusions

This section has described how the MMIT system works and reports on an experiment that shows the system's usefulness in improving performance in a simulated remote inspection task. The particular task was patterned after cleaning and inspecting a curved weld. The system demonstrated improved performance in terms of accuracy, decreased dependence on the quality of visual feedback to the human operator, and decreased variability between individual operators over more conventional approaches.

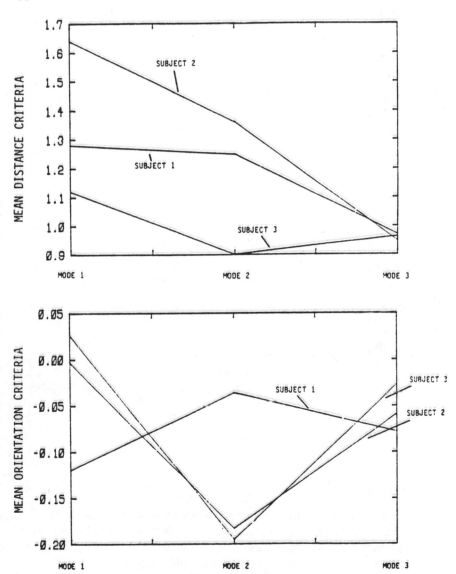

Figure 11. Mean performance as a function of control mode for the distance and orientation measures. No significant effects were seen.

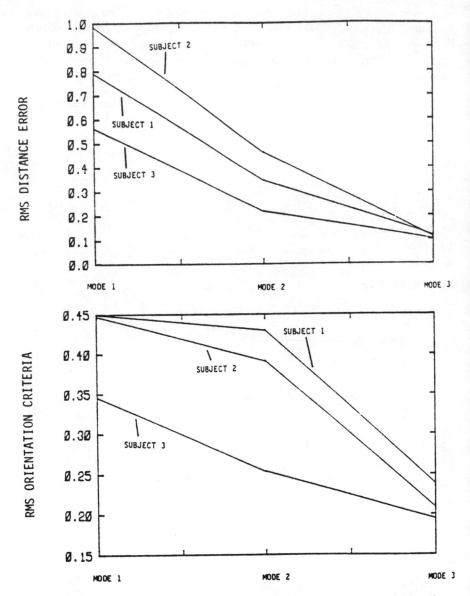

Figure 12. RMS performance as a function of control mode for the distance and orientation measures. The effect of control mode was significant for both measures. The variation between subjects was much less in mode 3 for both measures.

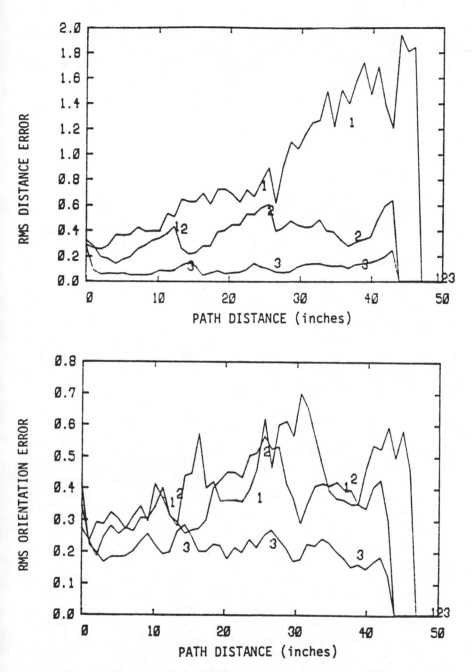

Figure 13. Performance as a function of distance along the weld. The upper plot is for the distance measure, and the lower plot is for the orientation measure. On each plot, curves for each mode are shown.

81

The computational and sensing requirements of the system are quite low. The system requires standard 16 bit microcomputers (LSI 11/23 for example) and needs only joint position sensing of the manipulator, making it practical for implementation on remotely operated vehicles. As much as possible, the system was designed to be manipulator independent.

C.   Factors Affecting Pointing: Telling The Computer About Geometry[4]

1.   Introduction

This work looks at the factors affecting the accuracy of "pointing" toward some prespecfied direction in space. To be more specific, accuracy in orienting the manipulator's arm normally or at a specific angle to some plane at some prespecified location are important for some teleoperation tasks. Both Brooks' (1979) and Yoerger's (1982) work showed that a complex trajectory may be defined by a computer if the cartesian coordinates of a number of anchoring points along the trajectory can be provided by the operator. To provide these coordinates, the operator has to locate the manipulator arm at each point in a prespecified (often normal) orientation. Thus, in this specific application, angular accuracy is of critical importance. In these experiments we investigated the operator's ability to orient the manipulator's arms normally to a plane at prespecified points differing in their spatial locations (part of Yoerger, 1982).

2.   Experiment 1: Direct Viewing, Varying Task Orientation

This experiment focused upon the ability to orient the manipulator's arm under direct viewing conditions. It is well known that manual motions performed on the mesial plane are more accurate than sidewise motions. To test if a similar relationship holds for angular accuracy, performance while orienting the manipulator normally to points on the mesial plane as well as on side planes was evaluated. We assumed that accuracy in different tasks should be rather insensitive to relative displacement of the viewpoint along the vertical axis. To test for this effect we used several planes differing in their inclination. And finally for each combination of horizontal displacements and inclinations five points differing in the relative x,y coordinates were evaluated.

Experimental Design—The Argonne E-2 master-slave force-feedback manipulator connected to a PDP 11/34 was used for the experiments. The experimental task was to orient the slave manipulator hand normal to a plane at a specified location on the plane. Three different planes were used. Each was at a different position, Figure 14 and each was inclined by a different angle, as shown in Figure 15. Each position was equidistant

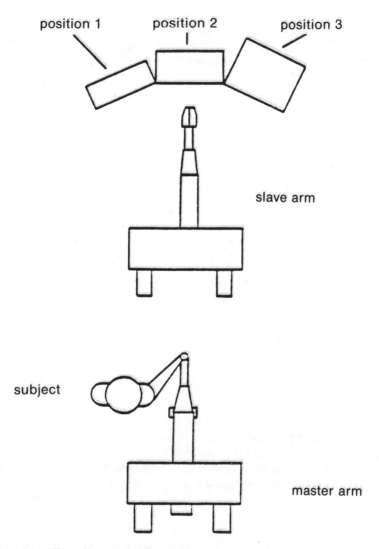

Figure 14. The direct viewing experiment: each plane could be placed in any of three positions. Each position was equidistant from the manipulator base. The center position was directly in front of the

from the base of the slave manipulator. On each plane, five locations were marked. Inclination of the plane, lateral position of the plane, and locations on the plane were taken to be the independent variables.

A $3 \times 3 \times 5$ full factorial within-subjects design was used. For this type of design, each subject performs the task for each combination of the

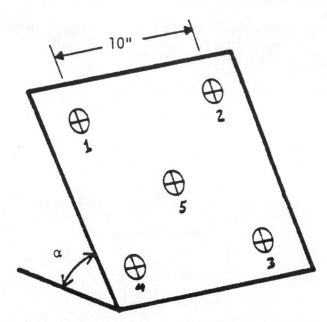

Figure 15.   Three different planes were used in the experiment. Each was inclined at a different angle α to the horizontal plane. Values of α used were 30, 45, and 60 degrees. On each plane, five locations were marked.

independent variables. The within-subjects design was chosen because it allows effects to be observed for a small number of subjects despite interaction between subjects and main effects.

Three subjects were tested, all right handed male engineering students with normal or corrected vision. Each subject performed three repetitions at each combination of inclination, position, and location, so that each subject performed a total of 135 trials in a randomized order. The trials for each subject were broken up into three blocks, with each block consisting of a fixed combination of inclinations and positions. The order of these blocks was counterbalanced across subjects.

Two components of angular error were analyzed which together describe the projection of a unit vector attached to the manipulator hand onto the task plane to be defined. Together, these two components give the magnitude and direction of misalignment. They will be called x error and y error. A perfect (perpendicular) alignment gives no projections. It is a point at the origin.

Procedure—Each subject was given the same written description of the task and how his performance would be judged. The instructions emphasized that accuracy was the prime performance measure, although performance time would also be recorded.

Each trial began with the master arm locked in computer control in the same position. The experimenter then told the subject at which plane and location on that plane the subject should position and orient the arm. The manipulator was then placed in manual control and the timer was started. The subject indicated when he had positioned and oriented the arm to his satisfaction by depressing a hand-held pushbutton switch. The timer was then stopped, and the current values of the manipulator's joint angles, the elapsed time, and the commanded position and location were recorded.

Results—For x error and y error, a method was devised for displaying both the mean and variance in a meaningful way. Examination of the data shows that the error does not vary independently in x and y. The variance of this two dimensional error is best described by a covariance matrix:

$$\bar{C} = \begin{vmatrix} \sigma_x{}^2 & \sigma_{xy} \\ \sigma_{yx} & \sigma_y{}^2 \end{vmatrix}$$

where the diagonal elements are the variances, and the off-diagonal elements are the covariances. In general, it is possible to find a set of coordinates for which the covariances are zero. The angle of the principal axes may be computed from the relationship:

$$\theta = \tfrac{1}{2} \tan^{-1} \left(\frac{2\sigma_{xy}}{\sigma_x{}^2 - \sigma_y{}^2} \right)$$

The values of the variances along these axes may be called the principal variances, uncorrelated measures of the spread of the data. The values of the principal variances may be found by the relations:

$$\begin{vmatrix} \sigma_x{}^2 \\ \sigma_y{}^2 \end{vmatrix}' = \begin{vmatrix} \cos\theta & \sin\theta \\ \sin\theta & \cos\theta \end{vmatrix} \begin{vmatrix} \sigma_x{}^2 \\ \sigma_y{}^2 \end{vmatrix}$$

The means and variability of the error data may be summarized by plotting an ellipse centered at the mean value of x and y error, with the major and minor axes of the ellipse equal to the square root of the principal variances (principal standard deviations). This plot provides a descriptive "error footprint" for the data x-y. (Figures 16 and 17).

Analysis of variance for x error showed significant effects for both position ($F(2,4) = 19.9$, $p < .01$) and location on plane ($F(4,8) = 11.9$, $p < .01$) while the effect of orientation was not significant.

A Newman-Keuls post-hoc test was used to test for significant differences between individual position means. This test showed that the mean x error was significantly different for each of the three positions.

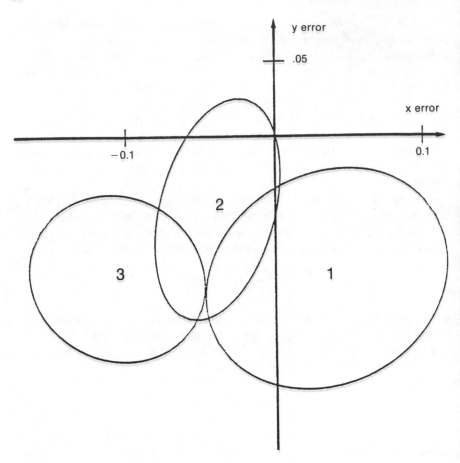

Figure 16. Orientation errors by position of plane. Positions 1 and 3 showed significantly higher y and radial error than position 2. For x error, all positions differed significantly.

The effect of position was significant for some y errors ($F(2,4) = 8.09$, $p < .05$). For y error, there was no significant difference between positions 1 and 3 (the left and right positions). Both positions 1 and 3 differed significantly from position 2 (the center position).

Interpretation of the error data for the position main effect can shed light on the relative importance of perceptual and motor considerations. The left and right positions, positions 1 and 3, are similar from a perceptual point of view, as the operator was positioned directly between these two positions. Position 2, the center position, was directly in front of the subjects, quite different, as the master slave manipulator is a right-handed

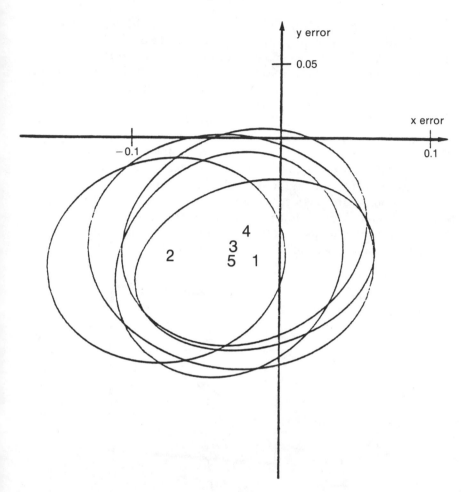

Figure 17. Errors by location on plane. No significant effect was seen.

device. All three means are shifted to the left due to common motor considerations.

3. *Experiment 2: Television Viewing Experiment, Varying Camera Viewpoint*

The previous experiment looked at the effect of position and orientation of the defined frame for a generally fixed viewpoint. In this experiment, one orientation was tested for different viewing angles. The task and experimental setup was the same as in the direct viewing experiment, but television viewing was used. Four different camera positions were used, as shown in Figure 18. Only one plane in a fixed position was used, as the previous experiment indicated that motor considerations are less important than perceptual factors. Again there were four locations marked on the plane. Camera position, location on the plane, and the practice effect were the independent variables.

Experimental Design—A full within-subjects factorial design was used. Eight subjects were tested. Each subject performed 8 blocks of 16 trials. Within each block, the subjects made 4 trials for each location on the plane in a different randomized order. After each block, the camera position was changed. The first four blocks corresponded to the first phase. In the first phase, the subject was given feedback about his performance from the graphic display after performing the trials for each block. In the second phase, the subjects performed blocks for each camera position again, but without feedback from the error display. Within each phase,

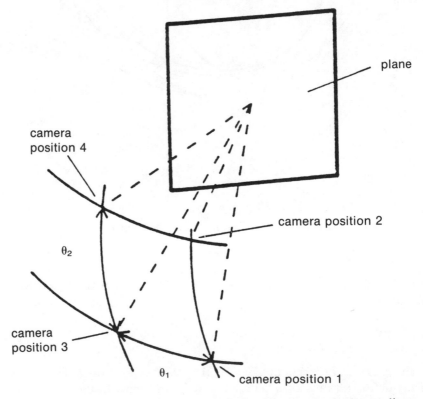

Figure 18. Camera positions were defined in a spherical coordinate system based at the center of the plane to be defined.

the order of camera positions was counterbalanced across subjects. Each subject was a right handed engineering student with normal or corrected vision. The dependent variables were elapsed time and orientation errors as defined in the previous experiment.

Procedure—Each subject was first given written instructions, emphasizing the same performance measures as in the direct viewing experiment. Subjects were also given written instructions about the meaning of the error display. Each trial proceded in the same manner as in the earlier experiment.

Results—Analysis of variance was performed on both x,y, and radial error and time data. Error plots were also produced, as was done in the earlier experiment.

Figure 19 shows the error data for the different camera positions. The difference was significant for y error ($F(3,21) = 41.$, $p < 0.001$). This difference in y error is shown more clearly in Figure 20.

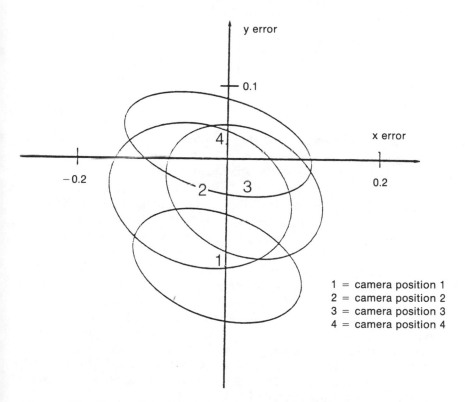

Figure 19. Error plots for different camera positions. The effect was significant for y error, but not for x error.

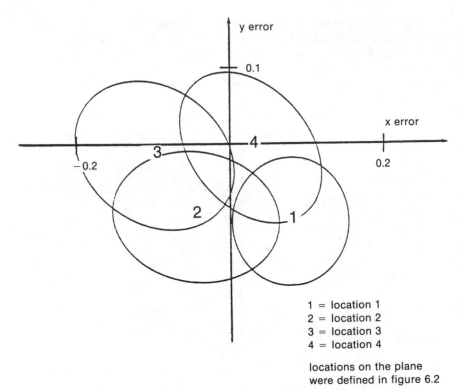

Figure 20. The effect of different locations on the plane can be seen in this plot. The effect of location on the plane on x, y, and radial error was significant.

Figure 20 shows the error plots for the different locations on the plane. This difference was significant y error (F(3,21) = 50., p < .001). Means for each combination of camera position and location on the plane were also tested. This interaction was barely statistically significant. Radial error, i.e., combined magnitude of the angular error, showed only small changes for each combination of camera position and location on plane.

Figure 21 compares the TV and direct viewing experiments. The mean and variation averaged across all independent variables and subjects for TV viewing is shown, along with the mean and variation for similar inclination of plane, and locations on plane for direct viewing.

The direct viewing experiment showed that this task is dominated by perceptual considerations. The television viewing experiment provides further clues about the source of these errors.

The significant differences in performance as a function of camera position confirm the perceptual nature of this task. For each location (i.e.

constant motor factors) y error varied significantly. The small interaction between location on plane and camera position shows that the location effect was fairly consistent for different camera positions.

The significant differences in performance for different locations on the plane can also be tied to perceptual effects and seem to support the idea that the direction of the error is determined by the viewpoint, as seen in the direct view experiment. The mean error for different locations on plane (Figure 20) is directly related to the actual positions on the plane. The errors for location 1 (upper left hand corner of the plane) are consistently toward the lower right hand corner, and so forth.

Examination of the plot of y error (Figure 22) shows several effects. Raising the camera always made y error more positive, while reaching for a lower location on the plane also made the y error more positive. This is the same bias toward the viewpoint which was seen in the direct

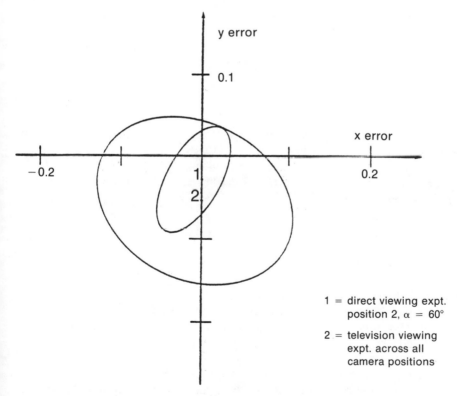

Figure 21. Comparison of direct viewing and television viewing experiment for similar inclination of plane, position of plane, and location on plane.

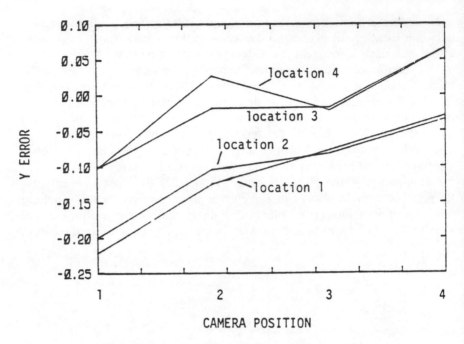

Figure 22. Y error for different camera positions.

viewing experiment. This corresponds to a consistent underestimation of the orientation of the plane relative to the direction of gaze, even when the direction of gaze is defined by a camera. This underestimation effect is consistent with the underestimation of radial direction constancy found by Hill (1976).

4. *Conclusions from Direct Viewing and Television Experiments*

The following conclusions can be drawn from these experiments:

1. While the average error with TV viewing was similar to the direct viewing experiment, the variation in error was larger.
2. The direction of error was consistently tied to perceptual issues corresponding to the spatial relationship between the plane to be defined and the direction of gaze.
3. For both direct and television viewing, subjects consistently underestimated the relative orientation between the direction of gaze and the plane to be defined.
4. A 45 degree angle between the direction of gaze and the plane to be defined was found to have best performance.

D.   Automatic Compensation For Motion Disturbances In Teleoperation[5]

1.   Introduction

This short section is included to give an example of a fairly straightforward form of partial automation or aiding which has been implemented in aircraft and spacecraft, power plants and other forms of human-supervised, semi-automatic systems. These are situations where relatively straightforward and conventional feedback control is employed to minimize disturbances which occur with respect to some variable while that same variable is being controlled in parallel or at a higher level by a human supervisor or machine.

Our example in this case occurs in remote manipulation of objects undersea where the manipulator base is sometimes a mobile submarine or vehicle which may move relative to the object being manipulated, and this makes either direct manual control or supervisory control difficult. This relative motion occurs either because a manipulator is being supported by a vehicle which is hard to hold steady against ocean currents or other disturbances, or because the object being manipulated is being buffeted, or both. The same problem could occur in space or in terrestial mechanical manipulation.

A means to overcome this is to make some measurement of the relative changes in displacement and orientation between manipulator base and object, either by optical, sonic or mechanical means, then to compensate for these changes by added motion of the end effector. The use of a mechanical "measurement arm" is one approach. Other means are optical, sonic, etc.

2.   Experiment

Hirabayashi (1981) implemented a measurement arm compensation scheme experimentally. He constructed a six degree-of-freedom (all angular movement) measurement arm which was lightweight and flaccid (offered little restraint). A six-degree-of-freedom Jacobian matrix transformation then allowed determination of the relative displacement of any object to which the measurement arm was attached.

Using a task-board with holes into which pegs were to be inserted, Hirabayashi drove the task board with a continuous random positioning device (three degrees of freedom, roughly 0.2 hz bandwidth, 6 inches root-mean-square amplitude). He then attached the measurement arm to this task board, and used the resulting measurement of displacement to produce a compensatory displacement bias between the master and slave.

When the arm was under computer control it compensated to within 0.2 inches, even with a crude three-foot-long measurement arm. Then computer compensation was added to manual master-slave control of actions relative to the moving object. It was found to be much easier with the compensation than without it to put pegs into the holes in the moving task board.

E.   Allocating Machines Versus Doing It Yourself[6]

1.   Introduction

This final set of experiments was relatively abstract in nature. We were interested in cases of man-machine systems where the human operator is given responsibility for optimizing system performance by combined allocation of *both* his own time and the time of various machine aids subordinate to him. All the tasks can be done manually if necessary, but time is constrained. Use of the machine aids may well enhance productivity. We considered only machine aids which could, once a task was assigned, complete it on their own. Eliminated, therefore, were machine aids like lawn mowers or pencil sharpeners which require continuous human control. This work was the thesis of Wood (1982).

In the specific experimental systems examined here a human subject (operator) is faced with a variety of tasks to accomplish (or task opportunities) presented to him on a computer graphic display. He is given a number of machine aids to supervise. That is, he can assign the aids to do the tasks or he can do them himself, in either case by pressing appropriate buttons on a key pad.

It was deemed reasonable to believe that as the operator makes decisions and assigns tasks he seeks to maximize some reward function. For simplicity, a undimensional reward function was assumed. It can be argued that in real life, people seek to maximize a variety of attributes like money, happiness, and respect. However, on a task basis it seemed an acceptable approximation to say that tactical decisions are based on maximizing a single objective.

Having hypothesized a unidimensional reward it is simplest to assume a linear utility for such reward. This assumption is crucial to allow exploration of the cognitive interface. In the experimental situation suboptimal performance by the subject should reflect some psychological barrier preventing the operator from fully grasping the complexities of the task environment. However, without a reasonably linear utility for the reward, it may be that the operator is trying to maximize a different reward function than that of the supposed optimal model. That is, the experimenter's optimum and the subject's optimum would not be the same.

2. Preliminary Experiment—BOXCLR

We assumed that the reward in our experiments should be "points," and that linear utility would be a reasonable assumption. A simple experiment performed at a computer-graphic terminal was devised to test this assumption. In the experiment, named BOXCLR, the operator was made aware at the outset of all tasks to be accomplished, all rewards, and all costs. By eliminating the subject's uncertainty and assuming he still knew how to optimize, we assume that any sub-optimal behavior was attributable to some non-linear utility for the reward, presumably because of some misconception about costs or rewards.

In BOXCLR, the subject was given a number of boxes or tasks to "clear" from the computer screen. This was done by assigning either a "human" (himself) or a "machine" (it looked like a bulldozer). The machine accomplished the task faster but cost more. Only one machine of each type could be operating at a time. When all boxes were cleared the experiment ended.

Three subjects each conducted 96 BOXCLR trials in which holding costs for a task not yet done were fixed, and where machine wages and the number of boxes varied. In this mini-experiment the subjects performed quite close to the optimal. A slight decline in performance was noted as machine wages increased. This may indicate that the operators did not have *perfectly* linear utility functions, but they were certainly close enough to justify use of the assumption in the experimental paradigm.

3. A Multi-Queue Allocation Experiment—SUPER

In this experimental paradigm subjects were faced with a dynamic multi-task environment where machine aids were available to assist the operator. To simulate the mental and physical separation of the operator from all but the task currently being performed, work areas were created and the subject could only look into one area at a time. This makes the experiment "multi-queue," with the operator searching from area to area for tasks. The experiment was performed as a simulation with information appearing before subjects on a computer screen "playing field." The computer display or diagram of the field is seen by the subject shown in Figure 23.

In the experimental game there are R classes of tasks that can possibly arise. For simplicity, only one member from each task class can appear at any time. Each task class arises in its specific work area, and only one area is displayed at a time. If a task exists in a work area, it is signified by the display of a box in the work area. To complete the task the operator must move the box to the right end of the area. The box may be moved manually by pressing the "DO TASK" button, or may be given to a

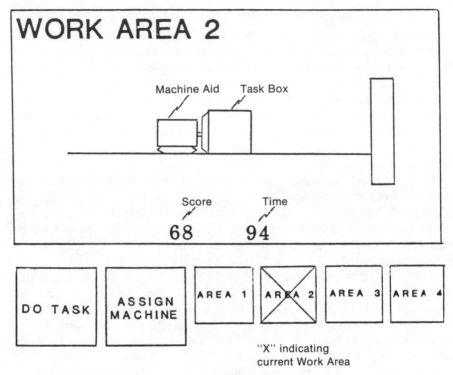

Figure 23.   Diagram of the SUPER playing field.

machine aid by pushing the "ASSIGN MACHINE" button. The operator also has the option of changing work areas. This is accomplished by pressing the control button for the desired new area.

A subject on SUPER faces only two types of decisions. If a task is before him, should it be performed manually or by machine? And, if there is no task, should he leave this work area to look at another? In making these decisions the subject must take into account all the variables which currently define the state of the system.

Three independent ratios using these parameters were found to characterize each task. These are:

- the ratio of reward to holding cost,
- the ratio of service rate of tasks to the arrival rate of tasks
- and the ratio of transition time between work areas to the mean arrival time of tasks.

The best strategy for two cases with the same values for these three ratios will be the same regardless of the absolute values of the task state variables.

4. Determining an Optimal Model for SUPER

It is fairly simple to determine an optimal model for operator behavior in a situation with very few variables. However, with increasing numbers of task classes, machine productivity levels and work areas, directly calculating the value of each combination of actions becomes too complex and some simplification must be used.

A decision tree maps out all possible paths open to the operator at each decision point. The reward accumulated for each unit of time is the most efficient method of evaluating a strategy's effectiveness. In simplifying calculations we sought a system which would provide an expected reward per unit time (RPT).

From the probabilities at each point in the tree the computer can determine the optimal strategy. However, the results from a decision tree which looks three steps into the future will be optimal only if the experiment is limited to three steps. To include more and more possible events in the analysis, the decision tree must be extended more and more steps into the future. In the limit, as the number of steps becomes infinite, the decision tree will incorporate all possible future development. Unfortunately, an infinite decision tree will have an infinite number of branches and the computation of expected RPT becomes impossible.

As the number of steps gets large, the incremental benefit of looking one more step into the future will decrease. It is possible to find a number of steps N that is sufficiently small to allow for computation of the "best" path, but which is sufficiently large to approximate the infinite tree. These approximately best paths or strategies are referred to as the "N-step optimal."

This model was validated and an appropriate number of steps was determined by running simulations. In cases where a low N model was found with the same effectiveness as a high N model the low one was chosen because of decreased computational time. In general, a six-step model was found to be most effective for determining the "best" strategies in this study.

5. Measuring Human Performance on SUPER

Human performance was measured by isolating each decision made by the subject and recording the game conditions at the time of the decision. The best-choice decision based on the optimal N-step strategy was then computed for the given conditions. The fraction of subjects' decisions which agreed with the computed best strategies was the resultant performance measure in this case. By analyzing each decision individually, separate performance measures can be computed for the different types of decisions. By comparing human and optimal performance measures it is possible to isolate the causes of human sub-optimality.

As suggested earlier, by employing a single dimension reward we expect to remove much of the nonlinearity and suboptimality in operator inferences. Another cause of sub-optimality, however, may be the operator's internal misrepresentation of the task parameters by the subject. While all task parameters were presented to the subject, in general he will accurately retain only some fraction of the information given. Some information he will forget and some he will make up to take the place of forgotten information. Unfortunately the link between the information presented and the information used by the subject to make decisions is not apparent.

Methods of presenting task parameters were varied in hopes that one would prove especially effective. Some improvement in retention was seen from relating parameters to stories and allowing subjects time to study them. Each subject was debriefed after the experiment to check his recall of parameters, strategy, and evaluation of performance relative to his strategy. Errors in recall were used as evidence of internal misrepresentation.

To try to further limit internal causes of sub-optimal behavior a task familiarization and training period was given each subject prior to the experiment. During this period the subject became accustomed to the computer display and his score was displayed as feedback for this practice time. The subject was exposed to only one task class at a time in this task familiarization period, thereby removing from consideration the decision to change work areas. In this way the subject was forced to concentrate on the relative merits of performing the current task by machine versus doing it manually. He was also forced to sit through the machine service time and time between task arrivals, items he might otherwise ignore while in the SUPER experimental situation. This practice served to bring subjects closer to steady-state learning. Once the experimental period was begun the score display was removed so that the subject could not use changes in the score (such as decreases due to holding costs from a task appearing in another work area) as a strategic indicator.

Subjects in SUPER experiments had three demands on their mental resources: they had to receive information from the SUPER display, they had to make decisions, and they had to implement these decisions. If the "rate" of an experiment, measured in terms of task arrivals and task completions per second, was increased it would seem that subject performance suffered because less time was available for decision making. A small series of experiments was conducted in order to determine a "fair" or "comfortable" experimental rate where subjects were not rushed in making decisions.

Three subjects were brought into the laboratory for three sessions each. In each session three experiments were conducted. These experiments were varied in the number of work areas used (called the scenario) and the experimental rate. Three scenarios employing two, four and eight machine aids were used, as well as three rates, fast, normal and slow.

It was expected that the "fast" rate would cause a decline in performance, but it should be noted that the "slow" time caused a decline as well. Subjects attributed this to boredom. They felt the excess time, and reported that long periods of inactivity made them forget what work areas they had viewed recently, thereby hindering their search strategy.

There were absolute differences in performance from scenario to scenario attributable to the differences in task environments. The two-work-area scenario showed better performance than that with four areas, because the search strategy required was less complex. The eight area scenario also showed good results because all tasks within that scenario were identical, so score depended on search strategy rather than on rewards and cost.

6. Experimental Results

Operators in SUPER faced two basic decisions—when to change work areas and whether to do tasks manually or assign machines. Our interest was chiefly concerned with the latter. Approximately seventy percent of all decisions faced by subjects, however, involved the question of whether to change work areas. A few general results regarding search strategy are therefore presented.

Transition Time. The importance of the decision to change work areas was dependent on the transition time. If the time was short, the effect of sub-optimal strategy was small because errors could be corrected rapidly. Operator decisions reflected this fact. When transition time was very short operators reported in debriefing that they only employed very simple search algorithms. When the time was lengthened, the subjects reported much more complex strategies.

Human Processing Limitations. A subject's search strategy depended less on the current condition of the experimental system than it did on the search pattern he chose to use. These search patterns or algorithms seldom took into account more than the last areas visited and where work was most recently found. Many simply pushed the work area controls in sequence, demonstrating how crucial the role of physical design may be. The exception to this was for work areas with very low arrival rates. Once these were discovered, subjects tended to skip these areas in their search patterns.

Machines That Are More Productive Than Their Supervisors. In assigning machines which could perform tasks faster than their supervisors subject performance was quite close to optimal with deviations attributable to search strategy. When the subject was given a machine that was highly productive at one task class but not at others, he assigned the machine there and performed the others himself. If the machine was efficient at all task classes, the subject would use it for virtually all the work.

Men and Machines with Comparable Abilities. Situations where men and machines are interchangeable are not uncommon. An airplane cockpit, as mentioned at the outset of this chapter, is one example. When subjects were given machines of ability similar to their own they performed quite well. A standard strategy was to search for a task and then assign a machine aid to it. If no machines were available the operator began the task himself until a machine became available. The N-step model generated an almost identical strategy.

Low Productivity Machines. In low productivity experiments subjects were given two machine aids, one with the same productivity as the operator, the second with productivity scaled down by a factor of X, with X ranging from 1 to 10. Subjects were told they did not have to use the less productive machine if they didn't want to. Yet, as demonstrated in Figure 24, subjects used the less productive machine far more often than the optimal model did.

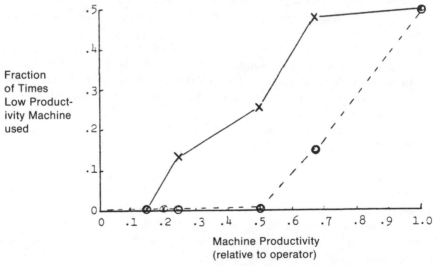

Figure 24. Fraction of all decisions to deal with a task where the low productivity machine was used—for decisions made by
— × —experimental subjects
— — 0 — — the optimal model

When questioned, subjects said they felt they would be wasting some of their resources if they didn't use the machine. A common evaluation of behavior was that if a low productivity machine was assigned in a work area not often visited, temporarily at least the number of areas to be searched was reduced. Subjects also felt it was better to have as many tasks as possible being worked on even if that meant using an inefficient machine.

The Effect of Machine Wages on Strategy.   Increasing machine wages should make a machine less desirable to use. Decreasing productivity should have a greater inhibiting effect because it not only spreads task rewards over a longer time but prevents new tasks and their potential rewards from appearing in the work area. This was not, however, borne out by the experimental results. Surprisingly, disinclination to use machines as machine wages increased was much more rational than in the case of decreasing productivity. However, as can be seen in Figure 25, subjects were not able to be quite as discriminating as the optimal model.

It is unclear why machine wages should have a greater impact on human behavior than productivity. It may be because in the task familiarization period the operator could see how fast his score started to drop when he used an expensive machine, i.e., negative feedback was more apparent and immediate. It may also be that subjects found cost a more tangible

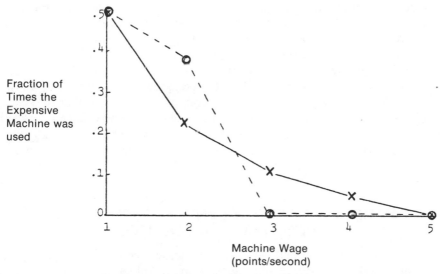

Figure 25.   Fraction of all decisions to deal with a task where the expensive machine was used—for decisions made by
— × —experimental subjects
− − 0 − − the optimal model

quantity than productivity. Unfortunately, machine wages are not one of the easily adjustable quantities in system design. Also, when an operator is an employee he is not the person who must pay the machine wage. Therefore machine wage may have only a very limited effect in the real world. When operators have to take on responsibility for the cost of their machines they may become much more inhibited than otherwise.

7. Conclusions from SUPER

A system designer concerned with the efficiency of an operator's use of machines should not be worried that humans will usurp jobs that should be performed mechanically. In fact, this study indicated that the opposite is likely; the operator will use the machine more than is optimal. Making operators aware of the true costs involved in machine use (maintenance, depreciation, capital investment, etc.) may decrease this tendency.

The experiments in this study dealt specifically with machine aids, and it is not clear whether the results are generalizable to the assignment of any tools, or to workers with variable productivities, or to tasks with changing requirements. Certainly machine productivity could be more deeply explored, especially in regard to the probability of machines successfully completing tasks. In this study machines did not fail; once a task was assigned to a machine it was always completed. In real life situations this is not always the case, and the effect of possible machine failure on operator performance is certainly an area for exploration.

Further, in this experiment the only requirement to put a machine to work was the push of a button. Real machines usually require set-up time, and this could be included in future studies.

Finally, our experiments used finite task queues. A new task could not appear in an already occupied work area. An implicit cost of leaving a task unattended in this study is that the rewards of possible new tasks in that area are foregone. By employing infinite queues where tasks "line up and wait" the costs of using a slow machine might be reduced and subject performance improved.

III. ILLUSTRATIVE EXPERIMENTS ON SENSING/DISPLAY AIDS TO THE SUPERVISOR

The experiments described below are concerned with the "affector" loop in Figure 2 (loops 1 and 2) and the "automatic" simulation/cognition aid, loop 8, which supplements the feedback from the remote process itself. Section III.A deals with the problem of constrained bandwidth in loops 1 and 2, one of the key reasons for having supervisory control of remote manipulators and vehicles. Section III.B deals explicitly with loop 8 from the viewpoint of supervisor simulation aiding. Section III.C tells of a special way to extend and enhance sensing and display by letting a low-level computer build up a visual image, thereby aiding the supervisor's memory (which method is curiously dependent on motor control, a kind of inverse of Section II.C. Section III.D gives an example of a well known display enhancement technique called a predictor display—one which promises to be very useful in supervisory control. Sections III.E and III.F deal with a special class of sensing/display aiding, but one which will assume greater and greater importance as automation increases, namely, that of aiding the supervisor in detecting and locating failures.

A. Bandwidth Limitation In Telemanipulation: The Framerate, Resolution, Grayscale Tradeoff[7]

1. Introduction

One reason for using supervisory control in space, in the deep ocean or indeed on terra firma is because the communication between human operator and a remote system is severely constrained, that is, the bandwidth is limited. For teleoperation in deep space one good reason is that radio energy is dissipated over the long distance to be spanned. If an electrical cable is used for deep ocean operations there need be no such problem, but there are problems of the tether becoming a large drag on the submersible vehicle and/or getting tangled up in structures that one wishes to inspect remotely. To avoid the latter problems one may employ acoustic communication. Even if a tether is dropped from a surface vessel down to within a few hundred feet of the submersible, acoustic transmission for the remainder of the distance can circumvent the problems cited above. However, because of sound energy dissipation this can only be done at the cost of having to reduce the bandwidth considerably relative to that for a wire (tether).

On terra firma as well as in space or undersea another bandwidth reducing factor may be that the operator may have to time-share his attention.

Thus one is left asking, for a given fixed communication bandwidth, how best to trade between the three variables of frame-rate (frames per second), resolution (pixels per frame) and grayscale (bits per pixel), the product of which is bandwidth (bits per second).

2. Experiments

These tradeoffs were studied by Ranadive [1979] in the context of master-slave manipulation. Experimental subjects were asked to perform two remote manipulation tasks using a video display as their only feedback while using our Argonne E-2 seven degree-of-freedom servo manipulator (in this case with force reflection turned off). Figure 26 illunstrates the experimental situation.

The first task was to locate a nut on a fixed bolt or knob and take it off by unscrewing it. (We abbreviate this task "TON" for take-off-nut). The second task was to pick up a cylinder and place it sequentially within the bounds of three fixed squares on the table which were numbered 1, 2, and 3, where the order of the placement, e.g. 3-1-2, was randomly drawn for each new trial. (We abbreviate this task "1-2-3".) Performance on each task was simply defined as the inverse of the time required to do that task correctly, and combined performance was the average of these inverse times.

The video display was systematically degraded with a special electronic device which allowed frame-rate to be adjusted to 28, 16, 8 or 4 frames per second, resolution to be adjusted to 128, 64, 32 or 16 pixels linear resolution, and grayscale to be adjusted to 4, 3, 2 or 1 bits per pixel (i.e., 16, 8, 4 or 2 levels of CRT intensity). Figure 27 shows the effect of resolution reduction.

Two subjects were used, both engineering students. They were trained for 10 hours in all combinations of display tasks and visual variables. When subjects first saw the video pictures with which they had to perform remote manipulation tasks, they refused to believe that they could succeed. Much to their surprise, however, they discovered that they were able to perform with a considerably degraded picture. During the data collection phase of the experiment subjects were allowed to practice on each display combination until "ready."

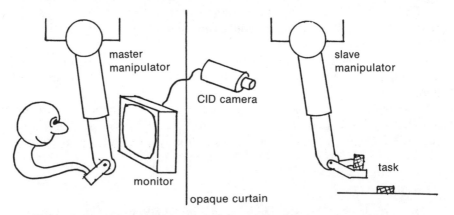

Figure 26.   Experimental configuration for Ranadive frame-rate, resolution, grayscale tradeoff experiments.

64x64

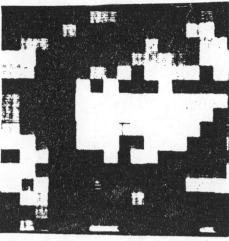

16x16

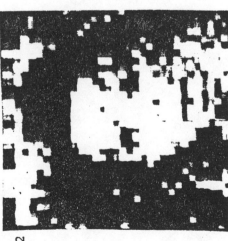

128 x 128

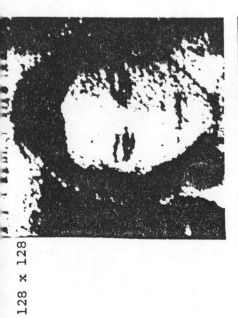

32x32

Figure 27. The same picture at various degrees of resolution (pixels).

105

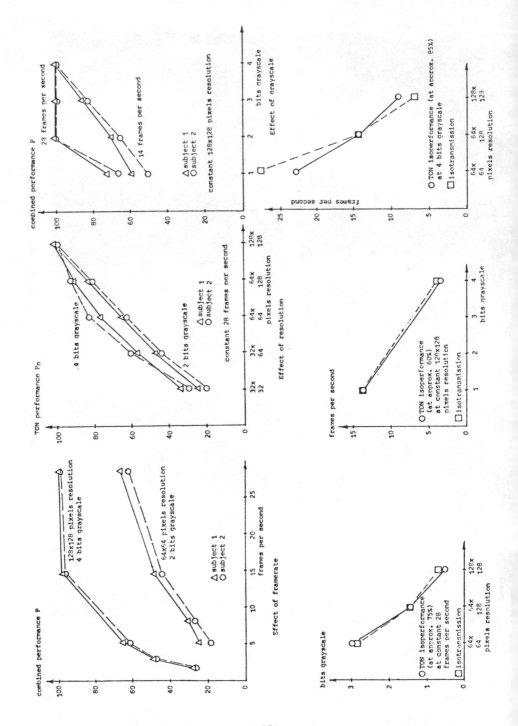

The data collection runs were ordered so that two of the three video variables were kept constant while the third was varied randomly among the levels for that variable. Ten times were collected (ten trials were run) for each combination (each data point).

Figure 28 shows the results. On the top row are shown the performance effects of frame-rate, resolution and grayscale while holding the other variables constant. Note that for frame-rate beyond 16 frames per second improvement depends on resolution and grayscale; performance improves smoothly for increases in resolution; for grayscale there is no improvement beyond 2 bits if the frame-rate is high enough.

On the bottom row constant level-of-performance tradeoffs (in this case using the TON task only) are shown for each of the three pairs of video variables. These iso-performance cruves (solid lines) are compared to iso-transmission lines, i.e., combinations of the two parameters which produce constant bits per second. It is seen that there is a remarkable correspondence. This means that for this experiment, and within the range of video variables employed, man-machine performance corresponds roughly to bits per second of the display, regardless of the particular combination of frame rate, resolution or grayscale.

Another result, though not tested systematically, was that subjectively much more noise appeared on each video picture at the slowest frame-rates than at faster frame-rates. It is believed that this was due to visual-psychological smoothing rather than anything electonic occurring at higher frame-rates.

Assume limited-bandwidth acoustic transmission is to be used as the means for communication between a human operator and an undersea teleoperator. It is clear that video will pose a far greater bandwidth requirement than other signals, probably more than all other signals combined. Therefore it seems reasonable to allocate fixed channels to other variables as required, then to reserve the bandwidth remainder to the combination of video signal requirements. That is, framerate, resolution and grayscale would *not* each have fixed bandwidth allocations; rather, provision would be made to trade-off between these as required, retaining their product as close as possible to the maximum.

To make this idea more understandable, assume that a given human operator of a teleoperator needs to get an accurate picture of a static object. He would like high resolution and sufficient grayscale, but frame-

Figure 28. Results from Ranadive experiment. The top row shows performance as one of the three variables is adjusted, the other two kept fixed. The bottom row shows isoperformance and isoinformation transmission tradeoffs for two variables at a time, the third held constant.

rate could be anything. In contrast, suppose the operator needed to monitor the speed at which a well known object moved against a background. Only enough resolution and grayscale would be necessary to get a good definition of what is object and what is background, but frame rate would have to be high. Either condition could be obtained by adjustment.

3. Operator-Adjustable F-R-G Tradeoff

Deghuee (1980) used an experimental computer-based aiding device which allowed the operator to make this three-way adjustment in-situ, i.e., he could modify the F-R-G (frame-rate vs. resolution vs. grayscale) tradeoff himself while performing a master-slave manipulation task of the type performed in Ranadive's experiments. In particular, the same master-slave manipulator was used with the force feedback turned off, and the same take-off-nut (TON) task, but a many-peg-removal task was chosen instead of the 1-2-3 task. Scoring was inverste time, the same as in Randive's experiments.

Four subjects were trained about ten hours each on each of the tasks. The same levels of the video variables were used as before. Two maximum bit rates (products of frames per sec, pixels per frame and bits per pixel) were used, one being 11500 bits per second, the other twice that or 23000 bits per second. The subject used three keys to call for any F-R-G combination, up to the maximum. When one factor was increased the other two were automatically decreased to keep the product at the constant maximum. Each subject, for every combination of task and maximum bit rate, performed both with and without the in-situ tradeoff adjustment capability. There were four trials for every cell of the experimental design. Data were analyzed by analysis of variance.

As might be expected, the use of the tradeoff control was significant ($p < .05$). Further, both the task main-effect and the task-subject interaction were significant ($p < .01$), a result not particularly surprising. What was more surprising was that the two maximum bit rates did not produce significantly different performance.

There was much variability in performance due simply to the fact that the visual interpretation time was extensive, and the real-time continual decision task of how to set the F-R-G combination added to this. It is believed that the means of making this adjustment could have been better "human engineered," and that this would have reduced variability and improve performance. Similarly the lighting was seen to be a critical factor, where amount of light affected grayscale adjustment and shadows provided important cues.

A principal result of this study was confirmation that with some training and some patience an operator can remove a nut with a remote manipulator using video of only 10^4 bits per second and with no force or tactile feedback. From the results an important special use of the adjustment became apparent in this case, namely to periodically but briefly increase resolution and grayscale at minimum frame rate in order to get confirmation that the peg was in the hole, or that another critical task phase had been achieved.

Use of this device is an important aspect of supervisory control, where the computer aid mediates the operator's instructions to provide, in this case, the best display (rather than control per se). This is loop 8 (of Figure 2) working in conjunction with loop 2.

B. Computer Graphic Manipulator Simulation for Planning, Training and Real-Time Feedback[8]

1. Introduction and Objectives

The objective of these experiments was to explore the development and use of a flexible computer-graphic simulation of a master-slave manipulator, a simulation which could be operated by a person in real-time and provide both visual and force feedback. Such a simulation would allow the operator, in effect, to see and feel his machine and work environment from any viewing or probing angle and at any scale, and to try out a variety of tasks, designs and control strategies without commitment to real hardware and associated costs and risks.

A realistic computer-graphic simulation can be of value both for experimentation and for operator training. Environments such as deep ocean work areas, which cannot easily be recreated physically in a laboratory, can be simulated on a computer instead. Further, by using the computer to control the simulated manipulator, it is possible to vary the kinematic and dynamic properties of the manipulator so that it can be used to simulate many types of manipulators. Several specific applications of the manipulator simulation are listed below (more details in Winey, 1980).

2. Testing of Control Systems

Building a control system for a remotely supervised vehicle can be both expensive and time consuming. With a new controller, there is always the risk of instability and failure which can result in damage to the hardware. When a control system is being developed for a one-of-a-kind prototype vehicle, the money and time lost in a failure can be disastrous. A computer simulated prototype controller can be changed quickly and easily, and failure generally involves little or no risk. A simulation can be run in and around instability in order to collect failure data without damage to hardware.

Supplementing of Operator Visual Feedback—A simulation may be especially helpful in remote work environments from which it is difficult to obtain pictures quickly or accurately. For example, as mentioned earlier, when unmanned vehicles are used in the deep ocean to avoid tethering problems they may be connected to the control system via an acoustic link. Such a link is subject to very limited (low bit rate) transmission; a television picture relayed this way must have either low resolution or low frame rate or both. The simulation, however, can be generated and updated in real time with only a small amount of data. All that must be known continuously is each angle of the manipulator's seven degrees-of-freedom. By superimposing a rapidly updated simulation of the manipulator on a slowly updated, but high resolution, television picture, the human operator is provided both the movement and resolution information that he needs.

Rehearsal—When an operator is required to perform a dangerous or delicate task in which a mistake could be harmful to himself, equipment or the task itself, it may be useful for him to practice the task on a simulator first. If a realistic simulation is available, the task can be rehearsed until the operator feels confident of his performance. If the manipulator is computer-controlled, the computer can monitor each of the practice runs and, when a satisfactory run has been achieved, the computer can be told to duplicate all or some part of it.

Humanizing Man-Computer Interaction—It addition to aiding the display of information from the work area to the operator, the computer graphic simulation can be used to accept information from the operator and to process it. The master manipulator can be used as a means of giving commands to the computer with the simulator serving to acknowledge understanding of the command.

The manipulator-plus-graphic-display can be used as a three-dimensional digitizer. Being capable of force-feedback it could assist the operator in locating reference points for such input, where some information, e.g., some attribute of position, is fed back to the operator as a force signal. The preceding are just a few examples of the practical applications of a computer graphic simulation of a remote manipulator.

3. Developing the Simulation

The major pieces of computing equipment used in the simulation were a PDP 11/34 computer running the RSX-11M operating system, and a Megatek 7000 Vector Display Processor with 3-D hardware rotate and 4096 lines resolution. The manipulator used as a master and simulated as a slave was the Argonne E-2 master-slave manipulator. It has seven degrees-of-freedom of motion and full bi-lateral force-reflection. Its electronical coupling allows it to be interfaced to the computer through an AN5400 A/D converter. Figure 29 shows the slave portion of the manipulator which was simulated.

The simulation was based upon simple but accurate representation of the manipulator arm and objects to be manipulated. These were stored in the computer using standard point-connectivity data. The manipulator arm was described in three separate pieces—the shoulder, forearm, and tongs—each treated as a separate entity. Each of the geometric elements was stored in an unrotated reference frame, and given a corresponding rotation matrix containing the transformations required to move that element from the reference frame to the desired location. Environmental objects for manipulation were further defined by "touching conditions" which described whether or not the object was in the manipulator tong's grasp.

Two types of objects and touching conditions were tested in the simulation experiments. The first object used was a sphere with a program specifying spherical touching conditions. When this proved successful, a stepped retangular peg was also tested. Two sets of rectangular touching conditions were defined, one for the main body of the peg and one for the stem.

Various dynamic and static properties for the objects, such as gravity, viscous drag, elasticity, and conservation of momentum, were included in the simulation and could be modified. This allowed great flexibility in simulating the environment within the computer. The manipulator and objects could be enclosed in a rectangular room. When a moving object collided with a wall of the room, it rebounded. This served to keep the objects within reach of the manipulator, as well as demonstrating conservation of momentum.

The simulation could be displayed from any viewpoint, and the viewpoint could be either stationary or in motion. The display could also be scrolled and zoomed so that any portion of the display could be observed in detail.

A simple submarine vehicle simulation was added, and made to be capable of same functions as the manipulator simulation. The manipulator was mounted on a vehicle which could be controlled in six degrees-of-freedom. The position of the vehicle was entered through an installed common block, which allowed the vehicle to be controlled by a secondary program.

Force Feedback—Two applications of force-feedback were introduced in the simulation. When a simulated object was gripped by the simulated manipulator, force-feedback was sent to the actual master. This was suf-

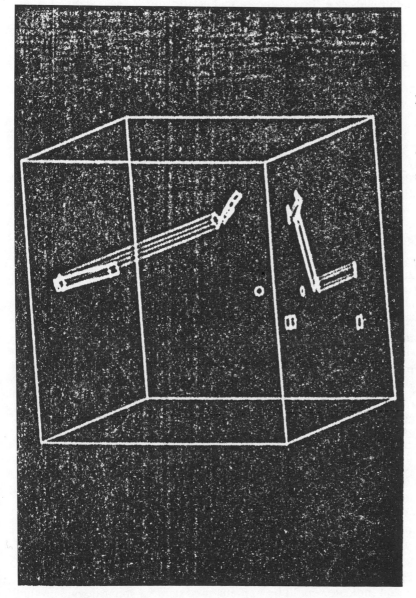

Figure 29. Wineys use of "shadows" and surrounding box to aid teleproprioception.

ficient to keep the simulated slave tongs open to the width of the object. The resulting sensation felt by the operator was that of an actual object within the tongs. The second application used force-feedback on all joints of the manipulator: A three-dimensional elastic surface was defined (Figure 30). The neutral surface (no force applied) was assumed to be relatively flat, eliminating the need to calculate the surface normal. Instead, the surface normal was assumed to be vertical.

Different elastic coefficients were assigned to various locations on the surface. Force-feedback to the master was generated proportional to the penetration distance beyond the neutral surface and the stiffness, so that the surface could be felt when it was touched. To provide a visual indication a gridwork approximation of the surface was displayed on the graphics terminal. To aid the operator in perceiving depth, the contour directly under the manipulator was displayed in darker linework. As the manipulator penetrated the surface, the contour deflected. If the surface was soft, deflection only occurred in the neighborhood of the penetration. If the surface was stiff, a larger portion of the surface deflected.

Depth Indicators—The major difficulty in using a video terminal for a manipulator simulation display is the lack of depth perception available to the viewer. When the simulation was first developed depth information was transmitted using traditional orthographic projections. This approach caused a few coordination problems and operators could become confused as to which view was front and which was side. As a result two other types of depth cues were tested to try to improve the operator's depth perception.

Experienced manipulator operators often rely heavily on shadows for depth information (Figure 29), so shadows with the source of illumination directly overhead were chosen as a second depth cue. The shadow was cast on an imaginary horizontal floor 50 inches below the manipulator's shoulder. Walls displayed on the screen were helpful in correctly orienting the shadow.

Unfortunately, to be understandable both the orthographic and shadow cues required extensive prior knowledge of the environment. The third depth cue display installed in the simulation, which did not require such environmental knowledge, was a proximity indicator which showed the absolute distance between the tongs and the object. This was displayed by a simple line on the terminal screen. The display was designed so that the length of the line was on the same scale as the display and was present any time the object was within 24 inches of the tongs. The indicator was ten times less sensitive at longer ranges. In real-life situations a proximity detector and indicator could be implemented using a sonar device on the tongs of the slave manipulator.

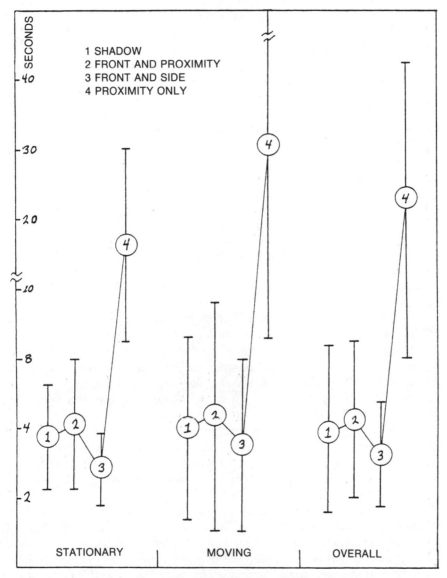

Figure 30. Evaluation of Depth Indicators.

4. Experimental Measurement of Operator Performance Using the Depth Displays

Three depth indicators, front plus side (orthographic projection), shadows, and absolute proximity in addition to the control were tested on five subjects. Two types of tasks were designed. The first required the operator to reach out and grasp a simulated two-inch, stationary sphere. The time it took the subject to grasp the sphere after the display was flashed on the screen was recorded. In the second task the conditions were the same except that the sphere was in motion. The sphere followed an orbit path about the surface of an ellipsoid. It was hoped that the moving task would highlight any coordination problems associated with the depth indicators.

The subjects were tested on a series of display types with both moving and stationary objects. The positions of the objects were selected so that they were always within reach of the manipulator and did not coincide with any real object which would obstruct the manipulator. The display types (depth indicators) were intermixed because subjects tended to become bored with repetitions of the same display. The order of the display was, however, kept constant, so subjects knew which type of display would be next. The positions of the sphere were arranged such that the average distance between successive positions was the same for each display type.

Four subjects each performed 80 to 90 repetitions of each combination of the four display types and two tasks. Two of the subjects had prior experience with the E-2 manipulator and the graphic simulation. The third had experience with only the manipulator, and the fourth had no prior experience. All subjects learned to use the displays in one to two hours. Learning curves were recorded for each subject to insure that performances plateaued before regular test trials began.

The analysis of how well each depth indicator performed is summarized in Figure 30. Clearly the proximity indicator with no additional display (the experimental control) gave poor results. So that it would not overwhelm the other three display types, the proximity indicator alone was left out of the statistical analysis.

A three-way analysis-of-variance was then performed. The results showed marginally significant differences between subjects, display types, and tasks. On both stationary and moving tasks, the front and side orthographic projections showed the best performance. Three of the four subjects said they preferred this display because it gave the clearest detail. All said it presented only slight coordination problems.

The use of shadows yielded the second best response times. All the subjects felt the shadow gave them the best perception of the object's position in the environment. The main difficulty with the shadow depth cue was that the manipulator's shadow tended to obscure the object's shadow when the two were in close proximity. Further, and although the result was not statistically significant, the shadow indicator seemed less affected by the motion of the object than by the front and side views. With larger or faster motions this could prove a significant advantage.

One subject preferred the front view of the manipulator with the proximity indicator because of its simplicity. His response times using this cue were close to those of the shadow. In order to obtain depth information from the proximity indicator the operator had to move the manipulator and watch the indicator's response. This caused the subjects little trouble, although occasionally they would search in the wrong direction initially. The proximity indicator is easily implemented in a simulation, but because it provides quite limited depth information it is probably not suitable for a complicated task.

Although there were differences in the effectiveness of the three depth indicators, these differences were small compared to the corresponding times required to perform the task. It is difficult to say with certainty which of the three indicators was best; each had advantages and disadvantages. Based on the experimental results, we suggest that the front and side views be combined with the shadow. The shadow provides an overall perception of the environment, while the side and front views elaborate the detail. An alternative in practical applications would be to allow the operator to select the view with which he is most comfortable.

5. Evaluation of Simulated Force-Feedback

One feature of the simulation which appeared quite promising but was not systematically evaluated was the use of actual force-feedback from the simulated forces. Force-feedback was generated when an object was grasped, giving the simulation a strong feeling of reality (Figure 31). Evidently the feedback was used by subjects in the simulation in two ways.

The first use of force-feedback was confirmational; it let the operator know when an object was grasped by the manipulator. If the object happened to slip from the tongs, the loss of tactile feedback let the operator know immediately.

The second use of feedback was quantitative. The force-feedback could convey information about the object, such as stiffness or weight. Surfaces ranging from extreme rigidity to a consistency approximating foam rubber were simulated. The main difficulty with simulating a variety of surface types came in the cycle time of the simulation. The softer the surface, the deeper the manipulator tongs would penetrate under a given applied force. The larger penetration motion required the manipulator to move

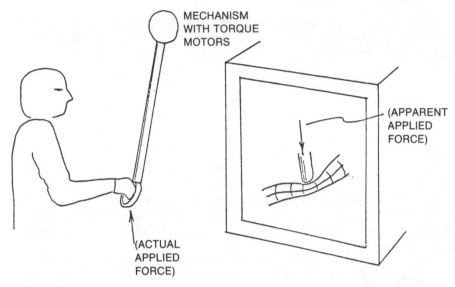

Figure 31.   Winey's experiment with touch (force feedback from a computer model.

greater distances on each program cycle. As a result, the simulation program had difficulty updating the manipulator quickly enough. Despite this problem, it was possible to simulate a wide range of surfaces with good response. Further testing and refinement of force-feedback would seem a promising area for further research.

6. Conclusions from Simulation Experiments

The manipulator simulation is believed to be a highly flexible and valuable tool for use with remote systems. The combination with environment and vehicle simulation also shows promise, allowing the operator to know the arm-vehicle-object positions in real-time instead of requiring the interpretation of pages of computer output. Further, the visual feedback is compatable with both manual and computer control.

Having a real human operator with an actual master manipulator interact with a simulated slave manipulator and environment or vehicle proved very workable and suggestive of various uses. Future research might look to improving the reality of the simulated environment or to using the simulation to rapidly evaluate alternative control strategies for alternative (future) environments. This would be particularly useful in-situ, when the environment is uncertain and the cost of making an error in the soon-to-be-encountered real environment is high.

Study of collisions between the object and other portions of the manipulator appears promising. For example, this might allow the object to be pushed by the manipulator without being gripped. The use of force-feedback could be expanded, allowing, for example, the object to feel heavy when picked up, or simulating a reaction force when the manipulator is struck by a moving object. An additional area of application is in use and evaluation of touch sensors, i.e., arrays of force-displacement sensors on the skin of the gripper which allow differentiation of force in space as well as magnitude and time. A third is to evaluate the video overlay of a rapidly updated arm simulation with an actual slow frame-rate but high resolution video picture.

C.   Blind Tactile Probing and the Inference of a Computer-Graphic Picture of Environmental Objects[9]

1.   Introduction

This section describes a novel means for tactile probing and discovery of the shape of an unknown object or environment. This technique offers promise for undersea operations where the water is so turbid that video is useless (and because high resolution acoustic imaging is as yet unavailable). It is the analog of a blind person probing in the dark by repeatedly touching at different points on an object or environmental surface in front of him and gradually building up a "mental image" of what is there, continually guiding his touching activity on the basis of what he discovers.

In performing "tele-touch" with a master-slave remote manipulator, if there were no dynamics and if force feedback were perfect it might be asserted that building up the necessary "mental image" would be no different than direct manual groping in a dark room. However every manipulator operator knows that is not reality; the master-slave manipulator itself is sufficiently cumbersome that one quickly loses track of where contact has recently been made and what the arm's trajectory has been. In performing tele-touch where a computer is determining the trajectory rather than a human operator's hand movements guiding a master, building up the "mental image" is still more difficult.

2.   Touch Probe Display

Flyer designed a unique touch-probe, a mechanical device which closes an electrical contact when it encounters a slight force from any direction. Then he programmed the 11/34 computer to determine and store the cartesian coordinates where any contact (touch) is made. He displayed on the Megatek screen, along with Winey's arm simulation, a projection from any viewpoint of cumulative touch points so stored. The operator can make no sense of such a display so long as the points are fixed. But the instant the image of points is rotated the shape and orientation of the one or more surfaces on which the contacts were established becomes immediately evident. What is a "mental image" in the case of direct manual grasping or touching becomes an explicit visual image. Figure 32 provides some (unfortunately static) examples of such displays.

As more points are added, the definition of the surface or object becomes more apparent. It helps somewhat to have the computer connect adjacent points with lines so that the best available "image" in three dimensions is a polyhedron and its planar projection is a polygon (or, if both front and back surfaces of an object are touched, two overlapping polygons). When rotation is effected the polyhedron immediately becomes evident. Rotation may be at a constant rate—usually around an axis near to or transecting the surface or object of interest—or may be controlled manually by a track-ball.

As contacts are made, points are added to the display, and what started out to be a polyhedron with few vertices and faces becomes a smooth surface, or a recognizeable object. The first few contacts between the manipulator probe and environment are made more or less at random. However, as the polyhedron takes on form, it is evident to the operator where to place the next few probes to provide the most discrimination and not waste effort and time by probing in the wrong places.

Another display trick Fyler demonstrated was to put the polyhedron into the (Lexidata) raster-graphic display generator's look-up table in such a way that the orientation of any facet of the polyhedron is determined. Then by use of the look-up table he "illuminated" different facets of the polyhedron on the raster display as a function of the orientation of each facet—as if the sun or light source were at one angle shining on a polyhedron (Figure 32). Again the operator was provided a trackball, in this case to let him move the apparent light source to any radial position surrounding the object, the polyhedron in this case being fixed in orientation, not rotating.

D.   Computer-Graphic Predictor Displays for Remote Vehicle Control With Transmission Delay and Slow Frame Rate[10]

1.   Introduction

Another form of computer-based display aid is the predictor display. This is a technique in which a computer model of the controlled vehicle or process is repetitively set to the present state of the actual system, including the present control input, then allowed to run in fast-time, say 100 times real-time, for some few seconds before it is updated with new initial conditions. During each fast-time "run," its response is traced out

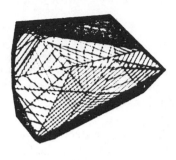

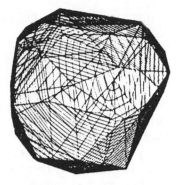

20 Points

50 Points

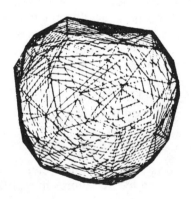

100 Points

Figure 32. Random touch points on a sphere which generate a polyhedron. As sphere is rotated the shape is easily perceived

in a display as a prediction of what will happen over the next time interval (say several minutes) "if I keep doing what I'm doing now." The general technique is about thirty years old, has been much discussed in the human factors literature (Kelley, 1968), and has been applied some to continuous control of ships and submarines. It still holds promise for a variety of future applications. It clearly has a role in supervisory control as a class of display aids.

2. Predictor Display for Remote Vehicle Control

When there is significant transmission delay (say more than 0.5 seconds) and slow frame-rate (say less than one frame per four seconds) a predictor display can be useful. Both of the latter conditions are likely to be present with long distance acoustic communication.

A random terrain was generated and displayed in perspective, updated every 8 seconds (Figure 33). A predictor symbol appeared on the terrain, continuously changing as the experimental subject controlled the motion of the vehicle, through a one-second time delay. Front-back velocity control was accomplished through corresponding position adjustment of a joystick, and turn rate by the left-right position of the joystick. Also superposed on the static terrain picture was a prediction of the viewpoint for the next static picture, and an outline of its field of view. This reduced the otherwise considerable confusion about how the static picture changed from one frame to the next, and served as a guide for keeping the vehicle within the available field of view. By use of the above *two* display symbols together, relative to the periodically updated static (but always out of date) terrain picture, subjects could maintain speed with essentially continuous control. By contrast, without the predictor they could only move extremely slowly without going unstable.

E. Computer-Graphic Simulation Aids for Failure Detection/ Location in Dynamic Processes[11]

1. Introduction

The Failure Detection and Location System (FDLS) is designed to aid the operator to detect and locate failures in real-time in systems such as: power plants (either fossil or nuclear), chemical plants, airplanes, ships, etc. In these systems power is transferred from available sources to locations where it is needed and converted into a desired effective rate of work. In the normal operational mode, the power transferred from subsystem i to subsystem j is P^n_{ij}. A deviation of the measured system power, P_{ij}, from its normal value P^n_{ij} (even though both may be changing with time) indicates that the system has failed. Hence, a system failure is defined to be a process which causes the transferred power, P_{ij}, to deviate from its normal value, P^n_{ij}. Examining P_{ij} and comparing it to P^n_{ij} is the basis of the FDLS method, developed initially by Tsach (1982).

It is desired to be able to detect system failures during steady state as well as transient operational modes. As a result, P^n_{ij} has to reflect the

PREDICTOR DISPLAY

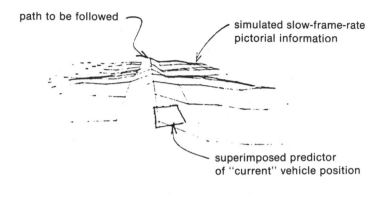

path to be followed

simulated slow-frame-rate
pictorial information

superimposed predictor
of "current" vehicle position

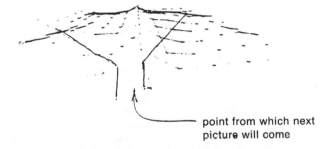

point from which next
picture will come

Figure 33. Simulation experiments with predictor displays (from Sheridan and Verplank, 1978). Slow-frame-rate pictures (8 seconds per frame) were simulated by computer-displayed terrain. The path to be followed was a ridge in the terrain. A moving predictor symbol (perspective square) was superposed on the static picture of terrain. The point from which the next picture was taken was indicated with a "table" (square with four legs) and the field of view was shown with dotted lines.

system dynamics. This can be done by constructing a dynamical model of the normal operation mode of the system. This model, implemented on a computer in real-time, will generate the desired reference power, P^n_{ij}, which is compared to the measured system power P_{ij}, as shown in Figure 34.

A simple comparison of real system to model turns out to be limited in its capabilities to detect and locate system failures. If the model is reliable, an unacceptable disparity between P_{ij} and P^n_{ij} indicates the sys-

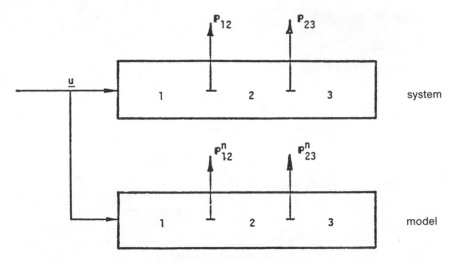

Figure 34. Comparing the system transferred power values, P_{12} and P_{23}, with their model values, P_{12}^n and P_{23}^n. u is a vector of the system's input variables.

tem has failed. Yet, this comparison does not indicate whether the cause is in the i-th, j-th or neither subsystem. Namely, a system failure is detectable by this means, but its cause cannot be located. Furthermore, in the case of multiple system failures, P_{ij} might not deviate from its normal value, P^n_{ij}, although the system has failed. This happens, for example, when multiple failure results in increased effort and decreased flow such that their product is unchanged. As a result, the power comparison test will not detect these kinds of multiple failures.

2. The Model-Based FDLS

To avoid the limitations of the system-model setup shown in Figure 34, one can modify this setup in a very special way. The power transferred from subsystem 1 to subsystem 2, P_{12}, is measured and its effort and flow variables are e_s and f_s. A model of each of the two subsystems (submodel 1 and submodel 2) is constructed. According to the causalities of these submodels, the measured system effort and flow variables are used as the model inputs.

In Figure 35 it is assumed that the causalities of the submodels are such that e_s and f_s are the input variables of submodels 1 and 2, respectively. The complementary variables, f_m and e_m, calculated by the model are the model output variables. Hence, the input to each submodel is a measured system variable, the covariable of which is calculated by the submodel.

The product of the input and the output of each submodel is the submodel calculated power. The power values of submodel 1 and 2 are:

$$P_{n1} = e_s f_m$$
$$P_{n2} = e_m f_s$$

By comparing these values with the system power P_{12}, system failures can be detected and the locations of their causes can be identified. If the model is reliable and P_{n1} differs from P_{12} the cause of the failure is located in subsystem 1, and if P_{n2} differs from P_{12} the cause is located in subsystem 2. A multiple system failure causes both P_{n1} and P_{n2} to differ from P_{12}. Thus the modified system-model setup shown in Figure 35 is capable of both detecting multiple system failures and locating their causes.

A comparison of the system and model power values is equivalent to the comparison of the corresponding effort or flow variables. The comparison of f_m and f_s (Figure 35) is equivalent to the comparison of P_{n1} and P_{12}, and e_m and e_s is equivalent to P_{n2} and P_{12}. In most applications the effort and flow comparison is advantageous. However, in electrical alternating current (AC) applications it is better to compare power. This is true since in AC applications both e and f avariables are continuously changing rapidly, whereas power is relatively constant.

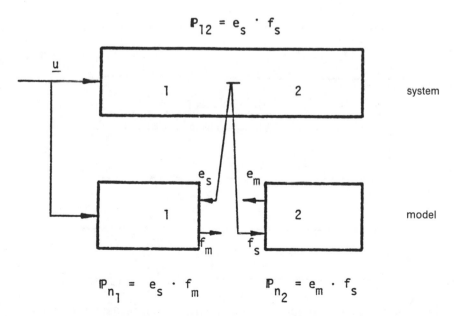

Figure 35. FDLM system-model setup where the system power, P_{12}, is compared with its model values, P_{n1} and P_{n2}.

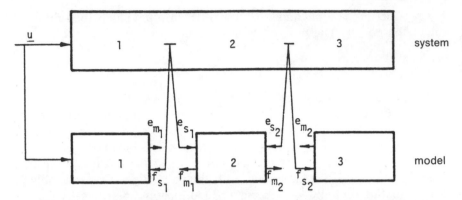

Figure 36. FDLM simultaneous mode. All the available system measured variables are utilized simultaneously and the model is disaggregated into small submodels.

In order to refine the location of the failure causes, one has to divide the system into smaller subsystems, and to perform a number of comparison tests as described in Figure 35. These tests can be performed either simultaneously (as shown in Figure 36) or sequentially (as shown in Figure 37). The trade-off between the simultaneous and sequential tests

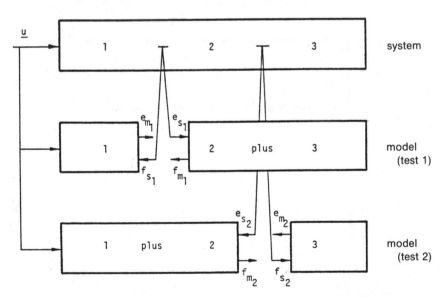

Figure 37. FDLM sequential mode. The system measured outputs of each point are utilized sequentially (test 1, test 2 . . .) to pinpoint the system failure.

is discussed in Tsach (1982). There a discussion appears on how to deal with the problem of performing the detection/location quickly, before the e and f variables have wandered too far out of their normal range and the submodels are no longer valid. It is also shown how to apply the FDLS to systems which are nonlinear, not-simply-connected, or thermofluidic.

The next section deals with how the measured discrepancies between actual system and model should be displayed to the human operator, and what data processing aids can be provided.

F.   Raw Versus Processed Data for Failure Detection/ Location[12]

1.   Introduction

One approach to detecting failures in dynamic systems is to run a computer model in parallel to the real system, as described in Section III.E above. In such cases the operator is expected to monitor the discrepancy and decide when it is sufficiently large to warrant concern and/or is *not* caused by known factors considered to be non-failures (e.g., equipment "locked out" for maintenance).

The question is then, how these variables should be presented to the operator. Should he observe the raw signals coming from real system and model, or should he observe processed information, where averaging and prejudgement is already made (suggested), or both? This experiment was set up to answer that question.

2.   Bayesian Decision Aiding

Assume that both the system and the model outputs have Gaussian distributed noise components. It follows distributed under normal operation, that D (the difference between the system and the model) is a Gaussian variable with $E(D)$ and variance σ_D^2, that can be estimated by taking measurement under nonfailure steady state conditions. An experienced operator, who has been monitoring a model system display, should be able to state the maximal model-system difference that might be expected under nonfailure conditions where he has no idea whatsoever which of the two hypotheses is correct, i.e., his prior distribution is diffuse. Now a sample of size n is taken and the mean difference d is computed. Since the sampled process is Gaussian, the posterior distribution is Gaussian also. Parameters are d and $\sigma_D^2/\sqrt{N}$. p($H1$) can be estimated as the area under the Gaussian curve prespecified by this hypothesis, and $p(H2)$ is simply 1-p($H1$). When the next sample is taken, the previous posterior distribution may serve as prior distribution which will result in combining the information from the two samples (see [Winkler and Hayes, 1970] for details).

Note that when much data indicating no failure has been accumulated, $p(H2)$ is almost zero. If a failure appears at this stage, much additional data may be required to overcome the certainty in the nonfailure hypothesis. In order to overcome this difficulty we decided to accumulate information only if $p(H2)$ exceeds 0.5, and to assume diffuse priors otherwise. This solution is similar to resetting the decision function to zero in Gai and Curry's (1976) model of the operator.

The effect of aiding the operator in information gathering by providing him with smoothed data and the effect of aiding him in the decision making stage were evaluated independently in an experiment described below. In particular, we compared performance in the case of a raw system-model display to the performance when either the model only or both the model and the system outputs were smoothed by averaging. Such analog information appeared either by itself or was coupled with some digital information related to failure probability. Under one condition this failure probability information was presented in terms of the failure/nonfailure odds, i.e. the ratio $\Omega = p(H2)/p(H1)$. We preferred this odds ratio over simple failure probability because of its higher sensitivity, and because evidence exists that in some tasks it results in less conservative behavior (Edwards, 1963). In another condition, whenever the probability of a failure exceeded 0.5 a "time flag," showing for how long this situation had lasted, was displayed.

An additional factor varied in the experiment was whether the compared variables (system and model) were state variables or not. Tsach (1982) has shown that when state variables (i.e., integrators) are used, the variables are automatically smooth and the resulting model of the outputs appears almost noiseless.

Under such conditions, the gathering and integrating of information can be done rather easily and, as a consequence, no effect of added smoothing should be expected when state variables (i.e., integrators) are used for comparison. However, it is still a question whether, under these almost perfect conditions for information gathering, information about failure probability will enhance performance. A positive answer to this question will help validate the two-stage information processing by the human operator implied in this work.

3. Experiment

Apparatus—A "prototype" of a hydraulic-mechanical-electrical system was implemented on a PDP-11 computer (see Figure 38). The simulation generated either the torque or the angular velocity of the generator's shaft as output variables. Gaussian noise with expectation zero and standard deviations of 0.55 of the measurement unit was added to the measured outputs.

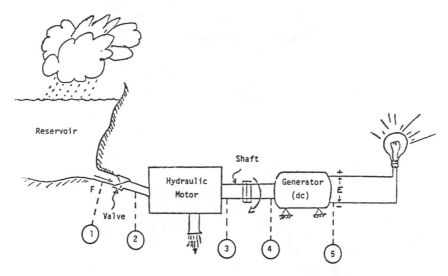

Figure 38. A description of an hydraulic-mechanical-electrical system. There are 5 measurement points available (Points 1–5).

A computerized model of the system was implemented in parallel to the prototype simulation. Failures were introduced by introducing a 10 percent disparity between the outputs of the prototype and its model. Because it was a state variable (i.e., follows an integrator in a dynamic sense) the model's torque was noiseless. However, when a failure was introduced, the model's torque diverged from its prototype only after 12-15 seconds, clearly a less desirable feature. On the other hand, both the prototype and its model angular veolcity were noisy, but in case of failure, their values diverged almost immediately.

Experimental Design—The analog information appeared in a running window centered on the Megatek 7000 Display monitor. The digital information appeared above the running window (see Figure 39).

Smoothing was achieved by connecting the means of samples of size 15 taken in equal time intervals each 1.5 seconds. The same samples were used for estimation of failure probabilities. The nonfailure hypothesis was defined independently for the "state" and "non-state" conditions.

The experiment was run in two sessions. Each experimental session was further divided into four blocks, one for each type of digital information. Each block contained 60 experimental trials, 20 in each mode of digital information, 10 of them with a failure. The times at which the failure could be seen were of course dependent on the time constants of the monitored variables and were longer in the state-variable-included condition. The trials in each block were randomly ordered, by using different random orders for different blocks. The order of blocks within sessions

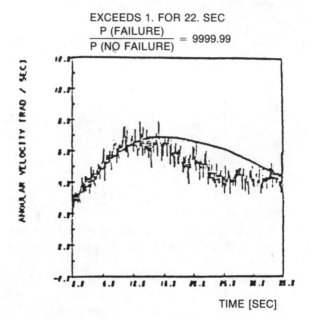

RAW SYSTEM & MODEL OUTPUTS

DIGITAL PROBABILITY RATIO AND THE TIME
DURING WHICH IT EXCEEDS 1

Figure 39.

and the order of trials within blocks were counterbalanced across subjects
by a Greco-Latin square. The order of sessions was balanced across subjects.

The experiment may be described as $2 \times 2 \times 3 \times 4$ factorial design;
the factor being Order of Session (state variable first vs. non-state variable
first), Type of Session (state variable included vs. not), Analog Information (raw, means for model only, means for both model and the system)
and Digital Information (none, odds ratio, time flag, time flag and odds
ratio).

Procedure—The subjects participating in the experiment were asked
to monitor in real time the system and model outputs, and by using all
the information available on the CRT display to decide if the system in
the monitored trial had failed. The different kinds of information that
might be available for decision and the definition of failure in terms of
the minimum system-model discrepancy were described and explained to

the subjects in the beginning of each session. The subjects were also told that the failure should be seen on the CRT almost immediately (no-state-variable condition) or only after 15 seconds (state-variable condition). They were asked to indicate failures by pressing an "alarm" button. No overt response was required for non-failure decisions. Accuracy and speed of response were said to be of equal importance.

A session began with 20 practice trials in which only analog information was displayed. Furthermore, each block was preceded by 10 practice trials. The practice trials were followed by accuracy feedback. No feedback was given in the experimental trails. Each session lasted for about 2.5 hours.

Before each trial, an empty window and a description of the information to be available in the trial to follow were displayed on the computer. Thus, the subject knew in advance what kind of analog (as well as digital) information to expect. Subjects started a trial by pressing a start button. The trial terminated when the subject "detected a failure," i.e., pressed the alarm button. When no failure was detected a trial lasted for 35 seconds.

Subjects—Eight paid volunteers, all of them students of engineering at MIT with some background in control theory, participated in the experiment.

Results—Accuracy data appear in Table 4. As can be seen there are almost no errors in the state-variable-included case. However, in the no-

Table 4. Percentages of False Alarms (FA) and Misses (MS).

Smoothing	Digital information	No Digital inf.		Ω		Time flag		time Ω & flag	
		FA	MS	FA	MS	FA	MS	FA	MS
No-state-variable included	Raw system & model outputs	13.75	11.25	0.0	0.0	3.75	0.0	0.0	0.0
	Raw system & smoothed model outputs	5.0	1.25	0.0	0.0	1.2	0.0	2.5	1.25
	Smoothed system & model outputs	0.0	0.0	0.0	0.0	0.0	0.0	0.0	0.0
State-variable included	Raw system & model outputs	0.0	0.0	1.25	0.0	0.0	0.0	0.0	0.0
	Raw system & smoothed model outputs	1.25	1.25	1.25	0.0	1.25	0.0	1.25	0.0
	Smoothed system & model outputs	2.5	0.0	0.0	0.0	2.5	0.0	2.5	0.0

state-variable-included condition, when both the system and the model were noisy and where the only available information was raw output, the proportion of error was higher, both in terms of misses and false alarms. As can be seen from Table 4 smoothing the outputs or providing the operator with failure probability estimations results in rather dramatic improvement of accuracy.

Mean detection times for correctly detecting failures were subjected to a four way analysis of variance with order of session as a between-subjects factor and type of session, analog information, and digital information as within-subjects factors. It was found that detection times under the state-variable-included condition were much longer than reaction times in the no-state-variable-included condition $[F(1,6) = 75.99, p < 0.001]$. Differences in digital as well as in analog information were both found to affect performance $[F(3,18) = 5.02, p < 0.02, F(2,12) = 9.97, p < 0.01$ respectively]. However, analog information interacted with type of session $[F(2,12) = 13.74, p < 0.001]$. Moreover, a second order interaction among analog information, type of session and order of sessions was also significant $[F(2,12) = 5.26, p < 0.025]$.

Newman-Keuls tests revealed that time flags did not result in hastening decision time over the no digital information conditions. However, the detection under each of these two conditions was found to be significantly slower than either in the odds ratio condition or in the odds ratio and time flag condition. The detection times in these two conditions did not differ significantly from each other. The above results are summarized in Figure 40.

The second order interaction between analog information, type of session, and order of sessions was further analyzed by testing the differences between the different types of analog information with the Newman-Keuls method for each combination of the two other interacting factors separately. As can be seen from Table 5 and as expected, smoothing had no

Table 5. Mean Detection Time (in seconds) as a Function of Session Type, Session Order, and Analog Information

	State-variable included			No state-variable included		
First session	Raw outputs	Model smoothed	Model & system smoothed	Raw outputs	Model smoothed	Model & system smoothed
State-variable included	16.76	17.15	16.83	8.00	8.97	7.54
No state-variable included	16.32	16.97	16.21	11.54	11.28	9.16

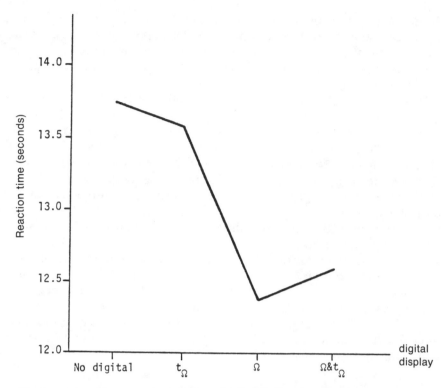

Figure 40. The operators' reaction times versus the different digital information displays.

effect in the state-variable-included condition, whereas otherwise smoothing of both the system and the model results in somewhat better performance.

4. Discussion And Conclusions

When the operator is required to detect system-model differences he is faced with problems of both information gathering and decision making. The main difficulty in information gathering is the distinction between the two outputs. It was found that such a distinction could be made easily when the model was basically smooth, i.e., when the comparison was made at state-variable points, just after integrators in the dynamics.

When both outputs were noisy, we could improve performance by smoothing the outputs of both the system and the model. However, there are good reasons for smoothing the model only. First of all, the operator should have the possibility to learn the noise characteristics of the system. Moreover, it may be important for the operator to know which of the two displayed outputs is the system and which is the model. Accordingly, we propose to leave the system unsmoothed. The efficiency of a smoothed-model raw-system display seems even more promising in the light of the expected improvement in dealing with noisy outputs as a function of practice, suggested by our data.

It was evident that the operators were aided at the decision making stage by providing them with the failure odds, which also are hypothesized to be helpful in detecting variance failures. However, in order to use such a device, the nonfailure hypothesis should be very carefully defined, and operators should be trained how to define failures in systems by providing them with feedback about the implications of their definitions.

We believe that simple digital information such as failure odds may be used efficiently not only for detecting failures but also to pinpoint the faulty component. In such cases, the operator may be required to monitor quite a large number of system-model comparisons. A good way to do it, without crowding the display too much, is to provide the operator with the failure odds ratio for each comparison. When the operator is alerted by a sudden increase in one of the displayed odds ratios, he can check his suspicions by getting the relevant analog information.

A failure detection and location system (FDLS) using a man machine interface designed along these lines is being developed in our laboratory and seems quite promising.

CONCLUSIONS

A number of experiments have been described to illustrate how supervisory control can be implemented in the context of controlling remote manipulators and dynamic systems, including detecting and locating failures. The experiments cited were clustered in two groups, the first relating primarily to command or "effectors" (human operator to computer to remote system), the second relating primarily to sensing or "affectors" (remote system to computer to human operator). In both cases the computer was providing essential mediation, much as a staff provides mediation between the top "boss" and the bottom level of workers in a large organization.

It becomes increasingly clear that supervisory control is not one function but many, and we may use further the analogy of functions served by middle-men in human organizations. When serving in the "effector" category the high level functionaries may concern themselves with translating the boss's orders into detailed instructions which are consistent with general policy (initial conditions) and recent orders, plus some forecasts (internal models) run by support staff. Lower level effector functionaries may receive these more specific orders and control the workers to put them into effect, making use of direct feedback from the workers. Such computer functions were embodied in the supervisory command systems described in Sections II.1 and II.2. Some other staffers may specialize in helping the boss formulate company policy by pointing out environmental facts (analogous to the "pointing" techniques in Section II.3). Some middle managers are assigned to nulling disturbances that arise (as in Section II.4). while some management advisors are concerned with how various personnel should allocate their time (as related to Section II.5).

In the "affector" category many of these staff mediators may be concerned about how and what to sample in the environment, much as economists, accountants and market researchers would (as with Section III.1 and III.3), whereas other mediators may use the sampled data plus a priori parametric information to run simulations of the situation to test and plan future actions (Sections III.2 and III.4). In some cases, the mediators may exercise such models as aids to detecting and diagnosing failures within company operations (Sections III.5 and III.6).

The main point is that in supervisory control the computer does not provide one function or one form of mediation, but many - at different places in the system, at different times or under different circumstances. This we believe will become evident with the various application of supervisory control—e.g., aircraft piloting and traffic control and office automation—as well as remote manipulation and process control. Nevertheless, though the computer takes on a diversity of functions, from the human operator's viewpoint the change to being a supervisor is always a change from continuous and direct sensing and control to indirect or somewhat remote control. The change means observing more integrated displays and issuing subgoal or conditional commands, all at a higher level than with continuous direct control.

The motivations for going to supervisory control are also many, as was implied in earlier discussion. They can be technical constraints on bandwidth, transmission time delay, or simple inability to provide sufficiently fast or accurate control signals in the direct manual control mode. Or it may be that the human operator is too busy or too fatigued or too bored to remain consistently in the control loop. In any such case the test of supervisory control is whether it works better.

It is proving difficult to determine what is "optimal" supervisory control. For one thing in supervisory control one of the operator's primary tasks is to set and modify subgoals and criteria; the "objective function" is not fixed. A second factor is the difficulty of modeling and experimenting on supervisory systems because of the inherent cost and complexity: it is not feasible to vary all parameters independently to find the best mix; there are simply too many parameters and the cost of changing is too great. Just the collection of data in the supervisory case proves difficult since the measures of human performance now have to do with how the operator communicated with the computer, what various displays he observed, what concatenations of commands he issued—each event likely to be different from the last. As supervisory control systems become more sophisticated it becomes less and less likely that any one simulus or response situation will be repeated.

So we seem to have to abandon our simple behavioral analytical models of the operator, where we have statistical confidence in few parameters by dint of many repetitions in well controlled situations, with few variables changing at a time. We feel a push to adapt a more holistic/synthetic approach to engineering man-machine systems.

In the pursuit of experimental control and clean straightforward science we cannot simply retreat to simpler manual control. Supervisory control is here, it works, it will be used and demanded to be made better.

NOTES

1. Section I—portions of this discussion are drawn from Sheridan (1982).
2. Section II.A—portions of this section are from Brooks and Sheridan (1980).
3. Section II.B—this section is based on Yoerger and Sheridan (1983) and Yoerger (1982).
4. Section II.C—this section draws upon material from Yoerger (1982) and Tzelgov, Yoerger and Sheridan (1983).
5. Section II.D—this work is based on Tani (1980) and Hirabayashi (1981).
6. Section II.E—this section is based on Wood and Sheridan (1983) and Wood (1982).
7. Section III.A—this section draws on Ranadive and Sheridan (1981), Ranadive (1979) and Deghuee (1980).
8. Section III.B—this section is based on Winey and Sheridan (1983) and Winey (1980).
9. Section III.C—this section is based on the work of Fyler (1981).
10. Section III.D—this work was first described in Sheridan and Verplank (1978).
11. Section III.E—this section is based on Tsach, Sheridan and Tzelgov (1982).
12. Section III.F—this section is predicated on Tzelgov, Tsach and Sheridan (1983) and Tsach (1982).

REFERENCES

Brooks, T. L., SUPERMAN: a System for supervisory manipulation and the study of human/computer interactions. SM Thesis, MIT Man-Machine Systems Laboratory Report, 1979.

Brooks, T. L. and Sheridan, T. B. Annual manual, experimental evaluation of the concept of supervisory manipulation. *Proc. 16th Annual Conference on Manual Control,* Cambridge, MA., 1980.

Deghuee, B. J. Operator—Adjustable frame-rate, resolution and gray scale tradeoff in fixed-bandwidth remote manipulation control. SM Thesis, MIT Man-Machine Systems Laboratory Report, 1980.

Edwards, W. Conservatism in human information processing. In B. Kleinmutz (Ed.), *Formal Representation in Human Judgement.* New York: wiley, 1968.

Fyler, D. Computer graphic representation of remote environment using position tactile sensors. SM Thesis, MIT Man-Machine Systems Laboratory Report, 1981.

Gai, E. and Curry, R. A model of the human observer in failure detection tasks. *IEEE Trans. System Man and Cybernetics* 6:85–94, 1976.

Hirabayashi, H. Supervisory manipulation for assembling mechanical parts while compensating for relative motion. SM Thesis, MIT Man-Machine Systems Laboratory Report, 1981.

McRuer, D. T., and Krendel, E. S. Dynamic response of human operators. WADC-TR-56-524, U.S. Air Force, 1957.

Ranadive, V. Video resolution, frame rate and gray scale tradeoffs under limited bandwidth for undersea teleoperation. SM Thesis, MIT Man-Machine Systems Laboratory Report, 1979.

Ranadive, V. and Sheridan, T. B. Video frame rate, resolution and gray scale tradeoffs for undersea telemanipulator control. Paper submitted to *IEEE Trans. Systems Man and Cybernetics,* 1981.

Sheridan, T. B. Supervisory control: Problems theory and experiment for application to human-computer interaction in undersea remote systems. MIT Man-Machine Systems Laboratory Report, 1982.

Sheridan, T. B., and Ferrell, W. R. *Man Machine Systems.* Cambridge, MA: MIT Press, 1974.

Sheridan, T. B., and Verplank, W. L. Human and computer control of undersea teleoperators. MIT Man-Machine System Laboratory Report, 1978.

Tani, K. Supervisory control of remote manipulation with compensation for moving targets. MIT Man-Machine Systems Laboratory Report, 1980.

Tsach, U. Failure detection and location method (FDLM). Ph.D. Thesis, MIT Man-Machine Systems Laboratory Report, 1982.

Tsach, U., Sheridan, T. B., and Tzelgov, J. A new method for failure detection and location in complex dynamic systems. Proceedings 1982 American Control Conference, June, Arlington, VA., 1982.

Tzelgov, J., Tsach, U., and Sheridan, T. B. Effects of indicating failure odds and smoothed outputs on human failure detection in dynamic systems. Paper submitted to Human Factors, 1983.

Tzelgov, J., Yoerger, D., and Sheridan, T. B. Factors affecting the angular accuracy of pointing with a master slave manipulator. MIT Man-Machine Systems Laboratory Report, 1983.

Winey, C. M. Computer simulated visual and tactile feedback as an aid to manipulator and vehicle control. SM Thesis, MIT Man-Machine Systems Laboratory Report, 1981.

Winey, C. M. and Sheridan, T. B. Computer simulated feedback as an aid to manipulator and vehicle control. Paper submitted to *IEEE Trans. on Systems, Man and Cybernetics,* 1983.

Winkler, R. L. and Hayes, W. L. Statistics, Probability Inference and Decision. New York: Winston, 1970.

Wood, W. T. The use of machine aids in dynamic multi-task environments: A comparison of an optimal model to human behavior. MIT SM Thesis, MIT Man-Machine Systems Laboratory Report, 1982.

Yoerger, D. Supervisory control of underwater telemanipulators: design and experiment. Ph.D. Thesis, MIT Man-Machine Systems Laboratory Report, 1982.

Yoerger, D. and Sheridan, T. B. Supervisory control improves performance for underwater telemanipulators. Marine Technology Society Meeting on Remotely Operated Vehicles, March, San Diego, 1983.

STRATEGIES FOR STATE
IDENTIFICATION AND DIAGNOSIS IN
SUPERVISORY CONTROL TASKS, AND
DESIGN OF COMPUTER-BASED
SUPPORT SYSTEMS

Jens Rasmussen

ABSTRACT

A framework useful for understanding and designing supervisory control systems is presented. Topics discussed include cognitive task analysis, hierarchical system representation, descriptive and prescriptive strategies for diagnosis, and phases of the system design process. The proposed framework is used to provide an integrated perspective of these topics in terms of implications of trends towards increasingly complex systems and sophisticated computer-based information technology.

Advances in Man-Machine Systems Research, Volume 1, pages 139–193.
ISBN: 0-89232-404-X

I. INTRODUCTION

Two major trends in technological development have influenced the problems to be faced by designers of man-machine interfaces of major industrial systems. One is a general trend towards larger and more complex systems with centralized control. Very often large plants imply large concentration of energy and hazardous material and, therefore, maloperation can have serious consequences not only for the plant itself and its operating staff, but also for the environment and the general public. In spite of the introduction of automatic safety systems, industrial accidents do occur, and analyses after the fact typically find significant contributions from operator errors. Recently, a number of major incidents, such as at Three Mile Island, have caused general concern for the operators' role and efforts to support them during complex disturbances.

The other major trend is the rapid development of modern information technology based on inexpensive, but powerful computers. Computer-based control systems and man-machine communication interfaces will change the work situation of industrial system operators in several different ways. The allocation of tasks to operators and to automatic control systems will change. Automation does not remove humans from the system, it basically moves them from the immediate control of operation into higher level supervisory tasks and longer term maintenance and planning tasks. This means that automation affects the general work organization. In general, this also changes the requirements for operator training, partly because the task is changed towards supervisory decision making in potentially risky situations, but also because the requirements for know-how during such situations may not be supported by the skill developed during more normal operating periods. Furthermore, computer-based interface systems result in new ways of preprocessing measured data and of interactive decision making which can be matched to various situations and different operator tasks. This affects the design and administration of operating instructions and manuals.

Extensive use of computer-based information technology in this way affects several aspects of system design which are typically considered separately, by different persons and at different phases of system design. However, these aspects are so intimately related that it is difficult to envisage them being gradually and independently introduced. Introducing new technology which affects several basic factors in system design is a kind of multidimensional optimization process in a multipeaked landscape in which the different summits represent designs based on different basic concepts. Considering that the traditional one-sensor-one-indicator technology has been subject to designers' experiments with improvements for half a century, it may be assumed that an optimum has been found. This optimum probably is rather flat, due to the adaptability of human operators and the opportunities left them for adaptation in the traditional systems. An extensive use of information technology will hopefully create a new and higher peak, which very probably will be more narrow due to higher system complexity and more intricate human-machine interaction.

Unfortunately, this peak cannot be reached in an optimal way by cautious changes of the traditional design along one dimension at a time, but only by a jump in all dimensions simultaneously. Furthermore, we can only expect to hit the peak within a reasonable vicinity of the top if we know where to look for it from the outset, i.e. if we have a conceptual framework to guide the jump.

The topic of this chapter is a discussion in some detail of one aspect of such a framework related to analysis and description of operator tasks in supervisory control. Use of computer-based information technology to support operators' decision making in supervisory control necessarily implies an attempt to match the information processes of the computer to the mental decision processes of an operator. This does not mean, that computers should process information in the same way as humans would do. On the contrary, processes used by computers and humans will have to match their different resource characteristics. However, to support human supervisory control, the results of computer processing must be communicated at appropriate steps of the decision sequence and in a form which is compatible with the human decision strategy. Therefore, the designer has to predict, in one way or another, which decision strategy an operator will choose. If the designer succeeds in this prediction, a very effective man-machine cooperation may result; if not, the operator may be worse off with the new support than he was in the traditional system, hence the narrow peak.

In the sections of this chapter, the following aspects of this problem will be considered: First, a framework for *cognitive task analysis* will be discussed. Traditionally, task analysis is discussed in terms of the control actions required by the system. This may partly be due to a behavioristic point of view that only descriptions in terms of observable events are professionally acceptable; partly due to limited need for description of mental activities in relation to systems of moderate levels of automation and simple one-sensor-one-indicator technology. What is needed is a framework for cognitive task analysis which enables the description of a decision task in terms of the necessary information processes. This framework or model must represent operators' performance in various situations and at different levels of training. It then also can be used to characterize information processing by computers.

Next, this framework will be used to relate the decision task to the control requirements of the system during various situations in order to identify the content of the information processes required. In particular, the diagnostic task will be considered.

For system design, we are not necessarily looking for predictive models of the detailed information processes as they will emerge in a specific situation. Rather, we need predictive models of categories of information processes enabling the prediction of that category which will be activated by a particular interface configuration and its display formats. The model will then support the choice of an interface design which can activate that category of behavior which has limiting properties compatible with the functions allocated the human operator, and for which the design has been optimized.

Two aspects of such a categorical model of the information processes are discussed in detail with reference to the diagnostic task. First, the knowledge of the functional properties of the system which is necessary for making control decisions, is discussed in terms of an *abstraction hierarchy*, similar to the hierarchies used for organization of data bases for CAD/CAM (computer-aided design/manufacturing) systems. Next, the different *strategies* which can be used for state identification and diagnosis in systems control, are discussed, together with their requirements for system knowledge and processing capacity.

In order to predict the strategy an operator will choose in a specific situation, it is necessary to know his subjective goals and *performance criteria*. This aspect is discussed with reference to analysis of the performance of technicians in a real-life task, based on verbal protocols.

Finally, the implications of this framework for man-machine task allocation and interface design are discussed.

II.   A FRAMEWORK FOR COGNITIVE TASK ANALYSIS

Design of man-machine interface systems based on modern information technology should be based on a generic model of the information processes implied in the decisions to be taken by the control system. To be generally applicable this model must be expressed in terms independent of the specific system and its immediate control requirements.

For control of a physical system such as an industrial process plant, a normative model of the necessary decision phases can be suggested. This sequence has been developed from analysis of decision making in a power plant control room and includes the following phases: First, the decision maker has to *detect* the need for his intervention and he has to look around and to *observe* some important data to have direction for subsequent activities. He then has to analyze the evidence available in order to *identify* the present state of affairs, and to *evaluate* their possible consequences with reference to the established operational goals and common policies. Based on the evaluation, a *target state* into which the system should be transferred is chosen, and the *task* which the decision maker has to perform is selected from a review of the resources he has available to reach the target state; examples of such target states and related tasks will be discussed in the following pages. When the task has thus been identified, the proper *procedure*, how to do it, must be planned and *executed*.

It will be noticed that the nature of the information processes changes during the sequence. In the beginning it is an analysis of the situation to identify the problem, then a prediction and value judgement and, finally a selection and planning of the proper control actions. Depending upon the context in which the control decisions have to be made, in particular whether it is control of a physical system governed by causal physical laws, or a social system governed by human intentions, the breakdown of the decision task may result in different elementary phases. Those used here have been chosen in order to have a flexible framework for description of the decision making found in verbal protocols from actual operation of a physical system, i.e., an industrial process plant (Rasmussen, 1976).

The protocols were recorded in an attempt to analyze the information processes of operators during plant operation. The scene was the control room of a fossile fuel power plant during a start-up by a team of highly skilled operators. As it turned out, the protocols did not in general reflect the information processes of the operators, but a sequence of "states of knowledge" representing what the operator knew about the operating conditions of the plant and about his task performance. These states of knowledge appeared as standardized nodes between the information processes shown in Figure 1. The information processes themselves were only reflected by the protocols when the operators had to cope with less familiar situations. This is to be expected with a highly skilled team during a familiar task.

The existence of break points in the information processes at standardized nodes between subroutines like those of Figure 1, appears to be very efficient for a variety of reasons:

- The information processes of the different phases have different structures since they are related to analysis, evaluation and planning in various functional frameworks, i.e., they are based on different information processing models. Therefore, a standardized "state of knowledge" separating them serves for recoding, being output state of the preceding process and serving as input state for the following.
- Breakpoints at standardized nodes make it possible to generate a sequence for special situations by chaining subroutines of general applicability and using solutions from prior experience. This is ex-

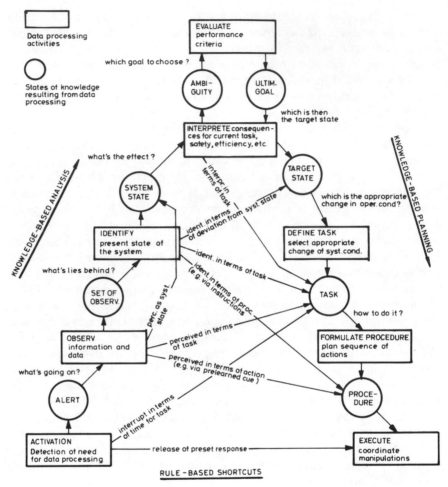

Figure 1. Schematic map of the sequence of information processes involved in a control decision. Rational, causal reasoning connects the "states of knowledge" in the basic sequence. Stereotyped processes can by-pass intermediate stages.

Source: Adopted from Rasmussen, 1976, by permission from Plenum Press.

tensively done by skilled operators leading to a great repertoire of short-cuts and by-passes in the decision process, as illustrated in Figure 1.

- Finally, standardized nodes are likely to be developed for easy communication among several operators sharing information and experience and cooperating in execution of the control tasks.

In addition to being a framework for representing the decision making of process plant operators, the "decision-ladder" of Figure 1 is well suited for man-machine interface design. The identification of the standardized nodes linking the elementary information processes, also makes it possible to plan in a systematic way the man-machine interaction in highly automated systems where system designers, operators and process computers will have to cooperate intimately in the control decision during complex disturbances. The background knowledge and data needed, vary considerably for the different phases of the decision sequence. So do the capacity requirements for memory and information processing. Therefore, the designer's prior analysis, the "on-line" decisions of the operator, and the support functions of the computer, will have different roles during the different phases. Proper allocation of decision tasks should be based on demand/resource matching, considering the requirements of the individual decision subroutines and the capabilities of the designer, operator and computer. It is, therefore, important to investigate the conditions under which the individual subroutines of the decision sequence can be formulated generically and separated from the overall decision sequence. In other words, it is important to find out whether the subroutines have some invariant structure across occurences. Since the identification or diagnosis of system state is a difficult task which in particular depends upon the actual conditions, this task is chosen for detailed discussion in the next section.

III. THE DIAGNOSTIC TASK

In order to formulate this problem in more detail we will consider the function of state identification in different situations during plant operation with reference to the decision-ladder of Figure 1. It is important to note that "system state" here is used in a broader sense than that of mathematical state space representations. It also includes the status of system configuration, such as "valving and switching states," as well as "failed states" of equipment.

The simplest situation is the identification of the states related to those control actions during the normal plant conditions which have been considered during design of the automatic control system. In such cases analysis has identified the relevant state/action associations which are then implemented in the control system. At the lowest levels in the control hierarchy this is done in the form of dedicated feedback loops where deviation between a measured variable and its set-point value leads to compensation through adjustment of a predetermined parameter. At higher level sequence control, predetermined patterns of plant variables are used as templates to test whether conditions for some planned actions on the plant are satisfied. In such cases, all higher level functions of the decision ladder have only been considered by the system designer; during on-line decision making, they are by-passed by stereotypical links between a pattern of observations and a set of actions. Similar state identification is used to protect the plant in case of disturbances and faults involving consequences that occur too rapidly or are too drastic in nature to be considered suitable for on-line human decision-making. Again, stereotypical links can be implemented in an automatic safety system acting on a predetermined pattern of state variables. An important feature of this state identification mode is that identification takes place with reference to data-pattern templates and relates directly to actions specific for each case.

The differentiated control actions necessary in response to faults and disturbances depend on more flexible identification of individual disturbances, and on-line decision making will include higher level processes. The extent to which this occurs, however, will depend on the actual circumstances. Consider first the state-identification or diagnosis involved in maintenance and repair. By definition, the target state of repair activities is the normal, physical configuration; the fault has to be removed by replacement of the damaged part. This task and the related procedure are well determined, and diagnosis simply means location of the faulty part, i.e. identification with reference to normal, physical state as illustrated in Figure 2.

Repair, however, cannot be the first control response to a fault during plant operation. For the initial control response, the target state is not given a priori. Even the goal, relative to which control strategies must be planned, will often be ambiguous—priority judgements must be made whether control actions should be chosen to maintain production, to protect equipment or people, or to minimize down-time in case of plant shutdown. In this situation, it is not possible to separate one well-defined diagnostic task from the overall decision sequence; the content of the task is not simply an identification of the primary cause in terms of equipment faults. Before attention is paid to this problem, the main concern will probably be to see whether the primary mass or energy balances have been disturbed and may lead to major risks. This means a search for deviations from normal in terms of major flow paths in the system. In order to evaluate resources for a compensation of such a disturbance, in terms for instance of heat sinks for improved cooling, the disturbance must be considered with reference to the actual anatomical configuration of the system and to possible reconfigurations.

This means that the diagnostic task implied in supervisory control, in the case of disturbances of an industrial process plant, may very likely require an iteration between consideration of the functional properties of

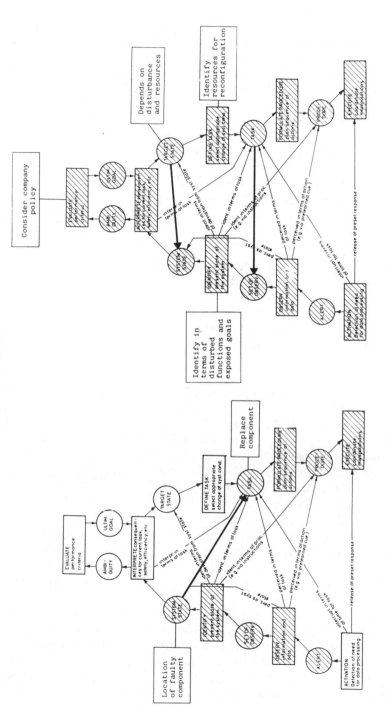

Figure 2. The diagnostic phase of a decision sequence can only be separated from its context in certain situations, like work-shop maintenance. In process control, the diagnostic phase will depend on goal priorities and available resources.

the system at various levels of abstraction. The implications of the actual plant state and of possible corrective actions must be evaluated against the overall operational goals and constraints; the functional relationships in the actual state should be considered with reference to the normal function; and the possible alternative resources in terms of equipment and supplies which can be used to counteract the disturbance must be reviewed. This process intrinsically is circular since the priority ranking of goals depends upon the nature of the disturbance and, at the same time, the level of state identification will depend on the goal being pursued.

The conclusion of this discussion is that in order to generalize a description of the diagnostic task, it is advantageous to consider two aspects separately. One is the search strategy which is applied for location of the disturbance. Another is the context, i.e., the level of description of the physical system, in which the search is performed.

Other phases of the decision process also depend on a description of the system at different levels. Counteracting a fault in the system during operation depends upon a redundancy in the purpose/function/equipment relationships of the system and an analysis of these relationships is also an important part of planning phases of supervisory decision making. This means that a systematic description of the system to be controlled in terms of a multi-level framework, an abstraction hierarchy, is an important basis for design of supervisory control systems.

IV. ABSTRACTION HIERARCHY

Description of man-made systems in terms of a multi-level abstraction hierarchy is well-known in design (Alexander, 1964). It has been used extensively for organizing data-bases for CAD/CAM, computer-aided design and manufacturing (Eastman, 1978). From a philosophical point of view, it has been discussed by Polanyi (1958, 1967).

In general, the description of a physical system can be varied in at least two ways. The description can be varied independently along the dimension abstract-concrete and the dimension whole-parts; the first dimension relates to the concepts used for description; the second, to the level of detail chosen for description. Later it will be seen that changes along the two dimensions are very often made simultaneously, but can in fact be done separately.

The number of levels of abstraction to consider depends on the kind of system in question and the purpose of the description. For representation of the functional properties of a physical system the levels shown in Figure 3 are found useful (Rasmussen, 1979). The space between the immediate physical appearance of the system and the functional purpose is bridged by five levels of description. Abstraction, in the present context does not

<u>LEVELS OF ABSTRACTION</u>

<u>FUNCTIONAL PURPOSE</u>

Production flow models,
system objectives

<u>ABSTRACT FUNCTION</u>

Causal structure, mass, energy &
information flow topology, etc.

<u>GENERALISED FUNCTIONS</u>

"Standard" functions & processes,
control loops, heat transfer, etc.

<u>PHYSICAL FUNCTIONS</u>

Electrical, mechanical, chemical
processes of components and
equipment

<u>PHYSICAL FORM</u>

Physical appearance and anatomy,
material & form, locations, etc.

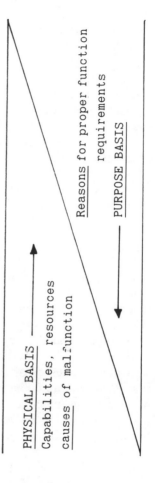

Figure 3. An abstraction hierarchy used for representation of
functional properties of a technical system.

mean simply removal of concrete detail. When moving from one level of
description to the next higher level, information representing the physical
implementation is discarded, but at the same time, information related to
the general co-function of elements is added which, for man-machine
systems, means information related to the purpose of the system, i.e.,
the reason for the actual configuration.

A. The Level of Physical Form

The lowest, most concrete level of abstraction is the representation of physical form, i.e., the physical appearance and configuration of the system and its parts. This level includes descriptions such as pictures, perspective drawings, maps of territories and locations of equipment as well as scale-models. Even at this level, the representation is not neutral or objective. The purpose of the system will very clearly control what is selected for representation, how the physical environment is structured in components and parts, and the resolution which has been chosen for description (e.g., a city map). Descriptions at this level are vital for interaction with a system, in order to find one's way, and to identify parts and components to manipulate. The relation to the higher levels is very often reflected in the labels and names attached to the physical items reflecting their functions and reason for presence in the physical map. Labels like pumps, valves, switches, and meters refer to the physical function of the components.

B. The Level of Physical Function

The level of description related to physical function represents the physical (i.e., the mechanical, electrical, or chemical) processes of the system or its parts. At this level, the description of functional properties is tightly related to the physical implementation which very often will be standardized technical components of widespread use. The description is focused on the specific physical equipment and the physical variables used to characterize its functional states. As examples consider: diesel engines with their processes of combustion cycles and mechanical transmission, transistors and their characteristic electrical relations in the form of signal matrices or graphic data sheets, centrifugal pumps with pump characteristics and limiting properties. This is the level of general professional knowledge about typical components and their properties. The properties of the physical items included in the descriptions very clearly are determined by their typical functions, i.e., the reason for their use, and their limiting properties. The level of detail, the resolution of the description, depends upon the task and profession of the people interacting. For a power plant staff, a diesel-electric generator probably will be a physical component, whereas an auto mechanic will consider an exhaust valve as a component. This is one of the reasons for keeping the level of abstraction and the level of decomposition separate in the formal framework.

The level of physical function is the level at which physically limiting properties are represented and at which causes of malfunctions are identified. Physical changes in components have functional consequences which propagate up through the levels of abstraction. The level of decomposition at which the fault is defined depends on circumstances. A power plant operator may locate the cause in a pump and from this make his decisions whereas the repairman has to identify the broken bearing seal. The description at this level of abstraction is based on concepts and relations characteristic of the physical process of the related components.

C. The Level of Generalized Function

At the next higher level of generalized function, the tie to the physical implementation is cut, and the concepts and language used for description are related to functional relationships which are found for a variety of physical configurations and which have general, but typical properties. Examples are feedback loops which have very typical properties, independent of their implementation in the form of mechanical, pneumatic, or electrical systems; properties of "cooling functions" can be established independent of the techniques used for pumps and heat exchange; the characteristics of an intermediate frequency amplifier can be stated in terms of gain and selectivity curves without reflecting the design in terms of tuned circuits or ceramic vibrators. Typically, the generalized functions represent the cofunction of equipment or components of different physical function and represent the functional structure of a system at a level of decomposition which is above the level of standard components. There is normally no one-to-one relationship between the physical function of a component and its potential role in a generic function. In general, there is a potential many-to-many mapping which represents the functional redundancy that is necessary for the freedom of operators to compensate the effect of disturbances: braking your car generally depends on the physical function of the brakes; if however they fade going down a mountain road, you make brake by means of the motor in low gear.

There is usually a clear distinction between the properties of a system which are represented at the levels of physical and generalized functions. Descriptions of physical functions are oriented towards the functioning of physical components and equipment, i.e., models are structured according to available components. Descriptions at the generalized level deal with functional relationships which are widely found independent of material manifestations, i.e., generalized functions are structured according to available models of causal relationships. Generalized functions typically are the subjects of theoretical studies in classical branches of engineering and descriptions are based on laws of nature, characteristic of the various professions such as Newton's laws in mechanics, Kirchoff's laws in electrical engineering, etc. and their specific practical derivations. Generalized functions are for instance power supply, heat transfer, feedback control, nuclear fission. Descriptions at the level of physical function are typically found in equipment manuals, component specification sheets and text books for technicians and equipment designers.

Descriptions at the level of physical functions clearly reflect the purpose of a system and the reason for the presence of the physical components. Description of a material item as a pump presupposes pipes and fluids and therefore its purpose. In the same way, descriptions of generalized functions reflect the co-function of various physical processes and, therefore, the reasons behind system design.

At the levels discussed so far, the functional properties of a system are represented in terms of the interaction among a set of typical processes, functions, or physical parts. To represent the overall function of a system by a consistent model, it may be necessary to move up in abstraction to a level which is independent of the actual physical and functional properties, but more tightly related to the intended, proper functional state of the system and its coupling to the environment.

D.   The Level of Abstract Function

At the level of abstract function, the overall function of a system can be represented by a generalized causal network, e.g., in terms of information, energy, or mass flow structures reflecting the intended operational state or, more specifically, in terms of flow of products, commodities or monetary value through the system. We are here in the domain of information theory and of the laws of conservation of energy and mass. The laws and symbols at this level form a consistent structure which is device and process independent and satisfied by the system design. Cassirer (1921) in his discussion of substance versus function, characterizes the concept of energy as follows: "Energy is able to institute an order among the totality of phenomena, because it itself is on the plane with none of them; because lacking concrete existence, energy only expresses a pure relation of mutual dependency." Correspondingly, a consistent formal model of the functional properties of a system in device and function independent terms can be developed in the form of a mass and energy flow map. Such a consistent model is very well suited for identification of control requirements and for automatic diagnosis (Lind, 1981, 1982).

A model at this level of overall functioning of a system only has meaning when considering a properly functioning system, since the actual "mutual dependency" is determined by the co-functioning of all the elements of the system, and upon the actual couplings to the environment such as inputs and power supplies. Therefore, representation at this level depends on knowledge about the purpose of the entire system and about the reasons for the actual structure.

The transition from the level of generalized function to that of abstract function is probably most evident when considering information processing systems. Here, a set of coding conventions relate the actual functioning of the system at the physical and generalized levels to the abstract function in terms of information processes. The abstract function represents the semantic content of the physical signals and hence the overall organizing principle. Consider for instance, the relation between the logic functions of an information process and the underlying generalized functions of a digital system, a relation which is based on a set of coding conventions. Also for energy producing or converting systems, the overall function of such systems is conveniently described at this level of abstraction in the form of the intended patterns of energy flows.

Abstract representation of organizing principles for complex systems is also well-known from the natural sciences for which they act as a kind of "reason" from which system properties can be derived: the first and second law of thermo-dynamics; the "survival value" of Darwin's theory; the principles of "least work" of Hamilton's theory; etc.

E.   The Level of System Purpose

At the highest level of abstraction, the purpose (i.e., the intended functional effect of the system upon its environment) is described. This can be done in terms of simple quantitative input-output specifications or in terms of the environmental, functional relationships of which the system is a part. Very often, however, the operational state is not totally specified and the ambiguities in the possible states of operation have to be removed by the control system by optimization from general criteria related to, for example, economy and conservation of resources.

For many kinds of systems, the ultimate purpose is related to the lower level of abstraction; for instance, the purpose of mechanical manufacturing plants, or treatment plants such as washing machines, clearly will be related to the level of physical function. Even there, however, the use of higher level descriptions may be necessary to consider system organization and performance, for instance with respect to energy consumption and resource conservation. Examples of the functional properties of specific systems, related to the various levels of abstraction, are shown in Figure 4.

F.   Formal Descriptions and Natural Language Descriptions

The functional properties of a physical system can be described at the various levels of abstraction in two basically different ways which are in several respects complementary. The description can be either in terms of a formal system of relations among variables or, in terms of a natural language description of objects and their properties and states and a de-

WASHING MACHINE	MANUFACTURING PLANT	COMPUTER SYSTEM
PURPOSE		

– Washing specifications
– Energy waste
  requirements
	– Market relations
	– Supply sources
	– Energy & waste constraints
	– Safety requirements
		– Decision flow graphs in
		  problem terms

ABSTRACT FUNCTION		

– Energy, water and
  detergent flow topology
	– Flow of energy and mass,
	  products, monetary values
	– Mass, energy balances
	– Information flow structure
	  in system and organization
		– Information flow
		– Operations in boolean logic
		  terms, truth-tables
		– Symbolic algebraic
		  functions and operations

GENERIC FUNCTION		

– Washing, draining,
  drying
– Heating, temperature
  control
	– Production, assembly
	  maintenance
	– Heat removal, combustion,
	  power supply
	– Feed-back loops
		– Memories and registers
		– Amplification, analog
		  integration and summation
		– Feed-back loops, power
		  supply

PHYSICAL FUNCTION		

– Mechanical drum drive
– Pump & valve function
– Electrical/gas heating
  circuit
	– Physical functioning of
	  equipment and machinery
	– Equipment specifications
	  and characteristics
	– Office and workshop activities
		– Electrical function of
		  circuitry
		– Mechanical function of
		  input-output equipment

PHYSICAL FORM		

– Configuration and
  weight, size
– "Style" and colour
	– Form, weight, colour of
	  parts and components
	– Their location and
	  anatomical relation
	– Building layout and appearance
		– Physical anatomy
		– Form and location of
		  components

Figure 4. Examples of descriptions in the abstraction hierarchy.

scription of events (i.e., changes of states, resulting from interactions among objects).

The description in terms of networks of relations among variables is the *deterministic* representation (Russell, 1913) used in theoretical analysis of systems in research and design and is typically implemented in the form of mathematical equations, tables or graphical curves. Formal rules for processing are typically available and the data used for processing are quantitative variables. The physical variables measured in physical systems by means of instruments are closely related to models in terms of variables and relations. Examples of such descriptions are at the level of generalized functions: differential equations of dynamic systems; pole-zero plots of feed-back loops; thermo-dynamic equations and Carnot-diagrams for heat transfer. At the level of physical function: transfer functions for transitors and related graphic representations; pump characteristics and specification of limiting properties.

Descriptions in terms of objects and their functional properties (i.e., how they interact and what they can be used for) are typical of the *causal models* used for common-sense, natural language reasoning. The two types of descriptions are in many respects complementary. In causal models, objects interact by events; the system is a set of objects related by a net of potential interactions in which changes or events propagate. Several quantitative variables are typically necessary to replace a description in terms of a state of a component or a mutual event between two components. The formal deterministic model is a network of relations among variables. In this model, the variables have replaced the physical objects as elements of the model. Physical objects are dissolved into a set of relations. To reflect a physical change of a component (due to a fault or physical damage) in a formal deterministic model, a complicated updating of a set of relations is necessary. Updating of the qualitative, causal model to reflect a fault is much simpler since the physical component is retained as an element in the model.

G.   Use of the Abstraction Hierarchy

At the lower levels of abstraction, elements in the description match the component configuration of the physical implementation. When moving from one level to the next higher level, the change in system properties represented is *not* merely removal of details of information on the physical or material properties. More fundamentally, information is added on higher level principles governing the co-function of the various elements at the lower level. In man-made systems, these higher level principles are naturally derived from the purpose of the system (i.e., from the *reasons* for the configurations at the level considered). Change of level of abstraction involves a shift in concepts and structure for representation as well as a change in information suitable to characterize the state of the function or operation at the various levels of abstraction. Thus an observer will ask different questions regarding the state of a physical system, depending upon the level of abstraction which is most suited to formulate the actual control task.

Models at low levels of abstraction are related to a specific physical world which can serve several purposes. Models at higher levels of abstraction are closely related to a specific purpose which can be met by several physical arrangements. The abstraction hierarchy, therefore, is useful for a systematic representation of the many-to-many mapping in the purpose/function/equipment relationship which is the context of supervisory decision making. For a process at any level of the hierarchy, information on proper function is obtained from the level above, and information about present limitations and available resources from the level below (Rasmussen and Lind, 1981).

In the present context, we are in particular interested in the human functions in man-machine systems which are related to correction of the effects of errors and faults. States can only be defined as errors or faults with reference to the intended functional purpose. *Causes of improper functions* depend upon changes in the physical or material world. Thus they are explained "bottom-up" in the levels of abstraction. In contrast, *reasons for proper function* are derived "top-down" from the functional purpose. This distinction is illustrated in Figure 3.

During plant operation, the task of the control system (which includes the automatic control system as well as human operators) will be, by

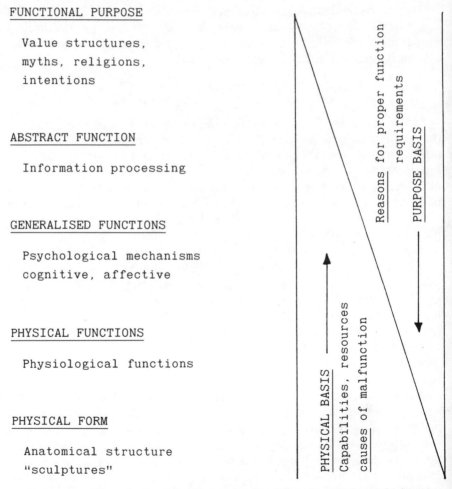

FUNCTIONAL PURPOSE

   Value structures,
   myths, religions,
   intentions

ABSTRACT FUNCTION

   Information processing

GENERALISED FUNCTIONS

   Psychological mechanisms
   cognitive, affective

PHYSICAL FUNCTIONS

   Physiological functions

PHYSICAL FORM

   Anatomical structure
   "sculptures"

Figure 5. The abstraction hierarchy used for description of human
functional properties.

proper actions on the system, to ensure that the actual state of the system matches the target state specified by the intended mode of operation. This task can be formulated at any of the different levels of abstraction. During plant start-up, for instance, the task moves bottom-up through the hierarchy. In order to have an orderly synthesis of the overall plant function during start-up, it is necessary to establish a number of autonomous functional units at one level before they can be connected to one unit at the next higher level. This definition of autonomous functional units at several levels is likewise important for orderly breakdown of system functions for shut-down and for reconfiguration during periods of malfunction. For such considerations there will typically be a tight coupling in the concrete/ abstract and in the part/whole dimension.

During emergency and major disturbances, an important control decision is to prioritize by selecting the level of abstraction at which the task should be considered. In general, the highest priority will be related to the highest level of abstraction: First, judge overall consequences of the disturbance for the plant production and safety in order to see whether the plant mode of operation should be switched to a more safer state (e.g., stand-by or emergency shut-down). Next, consider whether the situation can be counteracted by reconfiguration to use alternative functions and resources. This is a judgement at a lower level of functions and equipment. Finally, the root cause of the disturbance is sought to determine how it can be corrected. This involves a search at the level of physical functioning of parts and components. Generally, this search for the physical disturbance is of lowest priority. (In aviation: keep flying—don't look for the lost light bulb!)

When a disturbance has been identified and the control task located at a level of abstraction depending upon the situation, the supervisory control task includes the determination of the target state derived top-down for the chosen operating mode. In addition, the available resources for reconfiguration and limits of capabilities must be derived from levels below.

Another consideration should be added to this discussion. Frequently, other persons will be part of the environment which a particular person interacts with, and for which he has to use mental models in order to cope with unfamiliar situations. As for technical systems, various levels of abstraction can be used to model human "functional" properties, and an analogy of the levels discussed in Figure 3 is drawn for "models of man" in Figure 5. All the levels are used in various professional contexts, but what is of particular interest here is that, in ordinary working life, human interaction is based on a "top-down" prediction drawn from perceptions of other persons' intentions, motives, and on common sense representations of human capabilities, together with knowledge of accepted practice. Causal bottom-up arguments play literally no role and the most important information to use for planning human interactions for unfamiliar occasions therefore is knowledge of the value structures and myths of the work environment. The obvious reason for this is the complexity and autonomy of the human organism. However, it should be emphasized that, due to the growing complexity of information and control systems, the relevance of such "intentional models" (Dennett, 1971) is rapidly increasing also for interaction with technical systems.

In conclusion, the abstraction hierarchy serves the purpose of a systematic description of the context in which supervisory decisions are made. The other basic aspect to be considered is the information processing strategies humans can use for the different phases of the decision sequence and the performance criteria which control their choice in the actual situation. In the next section, these aspects of a diagnostic task are discussed from analysis of "real-life" performance, based on verbal protocols.

V. ANALYSIS OF STRATEGIES FOR DIAGNOSIS AND PERFORMANCE CRITERIA IN REAL-LIFE TASKS

In this section, information process related to diagnosis will be considered. The most complex data processing takes place when diagnosis is required during disturbances of system operation. Today, this task depends on the performance of human operators, but the trend is to use the data processing capacity of computers to support the task. To reach a proper man-machine computer cooperation, therefore, it will be necessary to study the diagnostic strategies which are actually used by operators in different situations. From this result, one can generalize and formulate a set of formal strategies which can be used as a basis for system design. Furthermore, it is very important to analyze the subjective preferences and performance criteria which guide the choice of strategy in a specific situation. Unless these criteria are known, it will not be possible to predict the strategy which a specific interface design will activate in an operator.

This implies the study of mental procedures used during real-life work conditions for which the use of interviews and verbal protocols are suitable tools. During the high season of behaviorism in psychology there were serious objections to the use of introspection and verbal statements. However, use of verbal protocols to identify data processing strategies during performance is not based on introspection; one is not asking the person to turn his attention away from the actual tasks towards his internal processes. One is just asking him to express his intentions, thoughts and needs (i.e., to externalize the internal verbalization which he may use anyway).

The quality of the protocols one can record, depends very much on the work conditions and the nature of the task. The diagnostic task aimed at repair has a much more well defined nature, than the diagnostic phase of supervisory control (see Section III). Furthermore, it is very difficult to obtain a reasonable number of good protocols from diagnosis in real-life plant operation due to the stochastic and infrequent occurrence of the related events. It was, therefore, decided to perform the basic study of diagnostic strategies in an electronic instrument repair shop (Rasmussen and Jensen, 1973, 1974). The usefulness of the generalized results was later confirmed by analysis of protocols from diagnosis in computer systems and process plants; together with analysis of error reports from power plants.

The situation in the electronic maintenance group of Riso National Laboratory was deemed to be very suitable, because there was a close personal contact and working relation between the people conducting the experiments and the repair men, and the natural interest of the maintenance group of a scientific institute to be directly involved in a research program related to its own professional methodology. The approach is discussed in detail in an appendix, because the method is important and the results proved to provide a good basis for generalizations.

A.   An Example

A discussion of the results of one analysis from a simplified example based upon the main features of one of the actual cases will be used to illustrate the approach. The case considered is one where a digital scaler displays two digits simultaneously in one decade, but otherwise functions normally. The task is now to obtain from the response of the system and by appropriate measurements, a reference to the location of the fault component.

In the present case, there is a close relation between the faulty parameter of the system response (i.e., the fault in display of the second decade) and a well-defined part of the system. The technician's interest will then quite naturally be limited to the circuitry connected with the second decade. Further reference to the location of the fault may now be obtained in different ways. It may be based upon detailed observations of the actual faulty response and consideration of the internal anatomy and functioning of the system. In the case under consideration, a design engineer localized the fault to a specific resistor in the decoder directly from the response, using his knowledge of the digital code and a diagram of the circuitry of the decoder. This method can, of course, also be chosen by a trained maintenance technician and, judging from textbooks for the training of such personnel, some authors consider it to be the "intelligent" method, i.e., to take few, carefully chosen measurements and use the observations in careful reasoning based upon functional understanding of the system.

Our observations indicate that a trained technician is most likely to choose another method, i.e., that of scanning through the faulty decade by a rapid sequence of good/bad checks of the actual signals against normal signals which are measured in one of the other decades, or which are found on the circuit diagram. In this way, the fault may be localized to the decoding circuit. In the circuit diagram this circuit is seen to contain less than half a dozen resistors. Therefore, rather than evaluate their function, it may be preferred to scan through the resistors by good/bad checks with an ohm-meter. Thus, an open circuit resistor is found.

B.   General Features of Diagnostic Performance

Although very simplified, this example illustrates some of the general features of the procedures found in the data collected:

- The basic feature of routines used may vary greatly in several respects. In the example given above, the designer used only a few observations, but employed complex data processing in his decision procedure. His procedure is related to the anatomy and internal functioning of the specific system and to the actual faulty condition. He treats several observations simultaneously, and his procedure is informationally economic.

    The trained technician uses many observations in a sequence of simple decisions. His method is a general search procedure which is not dependent upon the actual system or specific fault. He treats the observations individually in a stream of good/bad judgements which is informationally uneconomic, but fast.
- The technician defines his task primarily as a search to find where the faulty component is located in the system. He does not consider it to be a problem-solving task, which involves explaining why the system has the observed faulty response and understanding the actual functioning of the failed system.
- The procedures are organized as a search through a system which is viewed as a hierarchy of units. The system is composed of a number of subsystems: amplifiers, scalers, deflection generators, etc. Each subsystem has easily identifiable units such as amplifier stages, flip-flops, and oscillators, and these units have components, e.g., transistors, capacitors, and resistors.
- The general structure of the search can be broken down into a sequence of search routines, which are used to identify the appropriate

subsystem, state, or component. The structure illustrates how the technician attempts to sequentially limit his current field of attention. He is constantly asking the question of where to look next and thus tries to extract topographic references from his observations.

• A topographic reference from the observations is typically obtained in three different ways depending on very different depths in the consideration of the internal anatomy and functioning of the system.

Analysis of the protocols identified a number of such diagnostic subroutines which can be distinguished by the type of information which refers the technician towards the location of the fault. The strategies are discussed in detail in the following sections.

C.   The Functional Search

In the functional search, the topographic reference is obtained from the normal functional relation between a feature in the system response and a specific part of the system. A good example is troubleshooting in a TV-receiver. The man will scan the features of the picture in a stream of good/bad judgements and turn his interest to the subsystem related to the faulty feature. If the picture is too low, he will perform a search in the vertical deflection generator. The functional search is quite naturally the opening move in complex systems having subsystems with specific functions which are individually recognizable in the overall system response. However, the routine may also be used later in the procedure when the technician is faced with more complex data patterns such as wave forms on oscilloscopes.

The search is a special example of the topographic search which is discussed below in section V-D. The information pattern is scanned and familiar features are judged individually in a stream of good/bad judgements, and only results of judgements are normally used to control the next activity. If a response feature is judged faulty, attention is typically turned immediately towards the subsystem related to that function and a routine search is then performed in the subsystem. Information related to the observed mode of failure of the function is used in a very cursory manner.

In not a few cases, the technician studies the faulty response in great detail, but the information is not used to any large extent to control the search which follows. Sometimes the information is recalled later in the record to confirm a hypothesis found by routine search, and perhaps accompanied by a remark indicating a "eureka-feeling" when the hypothesis appears.

Faced with a multiparameter pattern of information, the technician normally has a good opportunity not only to deduce rather precisely what is the cause of the faulty response and where the fault is to be found, but also what sort of search procedure will be most efficient. However, one of the clear indications found in these experiments is that information available is not used efficiently in that way, even when faults were simulated to invite the use of short-cut methods by functional reasoning and evaluation.

D.   The Topographic Search

In the topographic search, reference to the location of the fault is obtained from the topographic location of a measuring point. The system is scanned by a sequence of measurements, and the observations are subject to simple, individual good/bad judgements.

The search is normally a test of performance along a main signal path in the relevant subsystem. The circuitry along the route is seen as a row of familiar units (e.g., amplifier stages) and by a sequence of rapid judgements the stage is localized in which the signal disappears or a faulty signal appears or an abnormal bias voltage is found.

When turning to a subsystem to perform a topographic search, normally very little information from previous observations (e.g. the nature of the fault) is carried over to assist the man in planning the search. The choice of the parameter to be used in the search is very dependent upon those norms for judgement which are immediately available. In some cases, the parameters chosen for the search cannot lead to the location of the fault, and information clearly indicating this may actually have been recorded by the man prior to the decision to start the search. However, the decision about where to look is often the only connection with the previous search. This may seem very inefficient, but still it should be remembered that a simple search of typically 5 to 10 measurements in a very rapid sequence may very often prove successful. Therefore, in the long run it may pay to take a chance and not consider every decision carefully.

In the records collected, it was found that the selection of the route and the steps of the search sequence are generally based upon the wiring diagram of the system. For this application, the diagram is not viewed as a functional description of the system, but solely as a *topographical map*, showing the information highways. Measuring points are chosen along the route of search at locations where convenient norms for judgements are available (e.g., where the diagram gives good reference data such as bias voltages or signal wave forms). The search is performed as a rapid scan along the route chosen and no judgements as to whether some of the steps will be informationally redundant seem to be made.

Literally speaking, all data collected during the search sequence are immediately judged good or bad individually. For these judgements the technician needs some model or description of the normal state of the system which supplies him with reference norms for each observation. The presence of a set of convenient reference norms is often directly stated in the record as the reason for the choice of the route and parameter used for the search.

If reference standards for the judgements are not otherwise available, the technician has to work them out himself by deduction from an understanding of the normal function of the circuitry, a task which is normally supported by the wiring diagram. If he also has to plan the route for search from a functional understanding of the circuit, he has to maintain simultaneously mental models at two different functional levels of the abstraction hierarchy, which is a considerable task. A model at one level is necessary to control the route of search; this model has to be related to the signal or information flow and thus to the function of the entire subsystem at the level of abstract function. Models at the other level of physical function are needed to supply reference norms for the individual judgements and thus have to be related to the detailed functioning of the sub-units along the search route. In that case, the procedure is slow and hesitating, probably owing to the considerable difficulty in simultaneously maintaining models at different levels.

If the necessary reference data for the judgements are hard to find, the measuring point or even the route may be discarded. Switching of attention to another route or field of search is often preferred to the effort needed to establish reference norms by functional reasoning.

E. Search by Evaluation

Search by evaluation of fault is used when the technician derives the topographic reference from the actual faulty response. This derivation implies an analysis of the information observed with respect to the specific instrument and its actual state of operation. This transformation may be illustrated as:

Observation (i.e., data observed describing the failed state) → Cause (i.e., what is changed in internal signals or functions) → Location (i.e., where is the faulty component resulting in the specific malfunction).

To be able to make such transformations, the man has to use mental models of the system relating changes in internal signals, parts, or components to the changes observed in system response. Clearly such transformations will be much more varied in their individual appearance than are the routine search procedures. Also the complexity of the transformations varies greatly from rapid statements based upon recognitions from previous cases to more complex deductions based upon several parameters and careful consideration of the internal system functioning and anatomy.

Such statements are generally expressed as recognitions. The transformations based upon conscious reasoning related to internal functioning of the system are complex and difficult to keep pace with during verbalization, and the records only indicate the surface of the activity. This fact, combined with the low number of cases, allows only very general and subjective attempts at classifying behaviour, but at least two groups of procedures seem to be used.

In one type, the technician seems to use a hypothesis-and-test search. He is working from inside the system outwards to the response in a way which could be illustrated as: Establish, by examination of diagrams or by memorizing, a mental model of the normal system anatomy, its signals and functioning. Then make a guess as to which signal or component might be involved in the faulty response. Modify the model accordingly and evaluate the resulting response pattern. Compare with the data observed to judge the relevance of the guess. This sort of procedure is most clearly expressed when the hypothesis is not a guess, but when the fault is found by another search procedure and the result is tested against system response by functional reasoning.

In other cases, the man is working from response data into the interior of the system. The procedure looks like a mental topographic search: From the response pattern and an understanding of the system, the absence of a normal signal or system state along a chosen search route is deduced by functional reasoning. The main difference from the normal topographic search is that the data, which are subject to individual judgement, are not measured directly, but deduced from the system response.

F.   Generality of Procedures and Depth of Supporting System Knowledge

The mental data handling necessary in a co-operative human-machine endeavour implies that the man has available some sort of mental model of the system and a procedure to use this model to process observed data. The mental model as well as the procedure may be supported by external means such as diagrams, drawings, instructions, and rules.

The previous discussion indicates that the mental procedures used in troubleshooting vary greatly with respect to the depth of system knowledge needed as support. At one extreme, technicians have procedures which are based only upon very general professional training and experience; at the other, they have procedures which call for very detailed knowledge of the specific system and the laws controlling its internal functioning. The experiment discussed here demonstrates the technicians' great ability to get around their search problem by means of a sequence of general procedures mostly depending upon their general professional experience and background.

In the preferred version of the topographic search, the search procedure consisting of a sequence of good/bad judgements is of very general applicability. The model of the system used for the search has only to supply an appropriate route for search and reference data for his judgements. If a circuit diagram is available which clearly indicates the main signal path and gives sufficient data for normal bias voltages and signals, the mental model needed has only to support the topographic correlation between the diagram and the system. The mental model need only be based upon professional experience with the visual appearance of typical components and circuits and with the normal layout of the circuitry.

The records also show a clear ability to base functional search upon very general mental models of the system. The technician will scan familiar features in the response by a sequence of good/bad judgements. If a faulty response feature is found, a general "block-diagram understanding" of the system can refer to the related subsystem. If this is a topographically well-defined part of the system, attention will immediately switch to this system in order to perform a routine search. Even when such faults were simulated so that the system response, as judged by the planners of the experiments, clearly indicated possible short-cut methods if the internal functioning of the system was considered, the technicians normally used their general search routines.

Also in the search by evaluation of the actual fault mode, the records show a pronounced preference for the use of transformation models which are not closely related to the specific system, but based upon general experience. Thus an interesting feature of the procedures found is the pronounced ability demonstrated by the technicians to produce overall procedures based on general search routines which are not closely related to the specific instrument. Scanning a high number of observations by simple procedures is clearly preferred to the preparation of specific procedures worked out by studying or memorizing the internal functioning of the system.

G.   Redundant Observations, Impulsive Decisions, and Mental Load

Other fundamental aspects of the procedure quite naturally follow the preference for general methods not closely related to the specific system or its actual fault. A general procedure cannot, of course, be based upon very detailed information found in the observations or measurements. In particular general methods cannot take advantage of information contained in specific relationships between several observations. This is clearly indicated by other features of the procedures found in the protocols.

The functional and the topographic searches appear as functional or topographic good/bad mappings of the system. Practically speaking, all observations are immediately judged to be good or bad, and only the results of the judgements normally control the next activity. In some cases, the parameters chosen for a topographical search cannot locate the fault, and information clearly indicating this may be recorded by the technician prior to his decision to turn to that particular search. Often, only the decision about where to look connects the routine to the previous search. During the search routines, no attention is paid to whether a measuring point will be informationally redundant or not.

The dependence of the procedures upon individual observations and judgements corresponds to a general tendency found in the protocols. Instead of making overall plans for the search, the tendency is to use rapid or impulsive decisions all along the search, based only upon the information observed at the moment. This, of course, gives a very individual pattern to the different overall procedures found in the individual cases. A main rule for the structuring of the procedures seems to be to follow "the way of least resistance." As soon as an indication is found that a familiar, general search routine may be applied, this is chosen without considering a possible, more efficient ad-hoc procedure. There seems to be a "point of no return" in the attention of the man the moment he takes such a decision, as discussed by Bartlett (1958). Although more information indicating possible short-cut methods or important hints for the next search is clearly available from the observation and is mentioned by the technician, the decision prevents any influence from such information; hence the next search is a routine, starting from scratch.

The basic difference in the amount of data needed for the different search strategies and in the complexity of the mental data handling task they impose upon the technician constitute important features of the various strategies available to the technicians. There is a complementary relation between these aspects of the routines. The very general procedures are based upon a rapid stream of good/bad judgements, and call for a large number of observations which are treated individually and then left behind. The system is mapped in a rather systematic way by such judgements, and this seems to be a convenient way of remembering the results of past activities. A general impression is that during his search the technician is well aware of his previous judgements. However, the originally observed data are discarded without subsequent recall, although in some cases, they seem to build up—unconsciously—a sort of "feeling;" and later in the procedure this feeling can initiate hypotheses that appear as "good ideas." The very specific procedures based upon system anatomy and functioning require only few observations, but the information handling is complex, and simultaneous treatment of several observations and a considerable carry-over of information may be needed in the short-term memory between the individual steps of the procedure.

This discussion focuses the attention upon the mental load on the technician during the task. As discussed thoroughly by Bruner et al. (1967), the mental procedures chosen by the technician may be strongly influenced by the constraints he meets in his limited capacity for short-term memory and inference. The multiple-task nature of troubleshooting may make this an important constraint. On a time-sharing basis, the technician has to formulate the route for search through the system by use of a diagram or by reasoning, locate the route in the real system, manipulate measuring devices, establish norms for his judgements from diagrams, experience, or functional reasoning, and keep track of his overall search.

Several indications of high cognitive strain are found in the data. A good example is a topographic search in a digital system performing logic operations, in cases when the technician has to plan the route and produce reference standards for judgements by deduction from an understanding of the functioning of the circuitry. As discussed earlier he then has to maintain simultaneously, mental models at different functional levels, and this is a considerable task. In this case, the procedure becomes slow and hesitating, and the person seems to be very insensitive to hints in the observation which would normally be familiar to him.

The records give several indications that difficulties in one of the subtasks tend to cause simpler procedures to be used in others. This should be taken into careful consideration when generalizing from clear-cut laboratory experiments with special equipment which eliminates all secondary subtasks.

H.   Fixations in Routine Search Procedures

The records indicate that the technicians have a great deal of confidence that the general search routines will ultimately lead them to the fault. If, for instance, a topographic search turns out to be unsuccessful and fails to result in a local search (which occurs in more than half of the attempts) the preferred decision is to repeat the search by another parameter. If this search, too, proves unsuccessful there is a pronounced tendency to return to a search performed earlier. This, however, seems to be a repetition with more careful judgements of the observations rather than a more careful evaluation of the actual faulty function. It should be stated that this is not necessarily due to inadequate ability to carry out functional reasoning, but more likely to the fact that these methods are inherently attractive as they consist of fast sequences, which are normally successful in the end. The behavior may be compared to that of most car drivers who prefer, when moving around in a big city, to drive along familiar main streets rather than preparing individual short-cut routes by means of a city map.

If the general search routines in special cases ultimately turn out to be unsuccessful, the technician often seems to "be in trouble." When in trouble, there is a tendency to rely on "good ideas" which admittedly seem to appear in most cases after a break or a period of confusion, which serve to break fixations. Such good ideas are difficult to trace. Sometimes, the technician returns to deviations met in a previous search, but passed over without further consideration; sometimes he expresses "a feeling that something is wrong around here"—a feeling which has grown from slight indications during earlier search sequences. In some cases important information has been recorded several times during routine search without triggering his attention until a period of confusion sets in.

When in trouble, there seems to be no tendency to consider it worthwhile studying by means of manuals or diagrams the functioning of the circuitry in greater detail. During a discussion of the strategies found in the experiment, the technicians stated that as a rule they found the general search routines successful. Apart from a "block-diagram understanding" of the system, it was not considered worthwhile studying the internal functioning of the circuitry. "If you run into trouble, better take a break, wait until the next day, or discuss the problem with a colleague." Asked if they could suggest types of cases for which they would find it worthwhile studying the internal functioning of the system in detail, the technicians said that it would be the case if measurements of manipulations could have serious consequences, as in live warning systems—"when a siren is at the end of the wire" or if the working conditions on site are unpleasant, e.g., owing to bad smells as in chemical plants. In other words, it would be the case if "costs" related to observation were high. A test case in connection with the level control system in a radioactive waste tank system resulted in a very "rational" procedure based upon a careful functional evaluation of the system response in advance and very few measurements on site.

It was also suggested to the technicians that they should use functional reasoning when in trouble in the normal repair shop environment. This, however, did not cause any significant changes in the procedures used in the records made thereafter. The procedures seem to be so highly trained that they are difficult to change by suggestion of "better procedures." When a troubleshooting task is running, a skilled technician seems to be completely absorbed in the task, and he does not "remember" the suggestion when difficulties arise. The test case with the waste tank system may indicate that, to change a procedure, the technician has in advance to perceive the task as one related to a special search strategy.

I. Subjective Formulation of Task and Performance Criteria

Important aspects to consider are the subjective formulation of the task and the performance criteria in the choice among the various search strategies available. This aspect is especially important since the formulation of the trained technicians may be basically different from that of a design engineer, who is after all very often responsible for preparing the working conditions for technicians in the form of layouts of systems, instructions, and operating manuals and diagrams.

The experiment discussed here clearly indicates that the task is defined by the technicians primarily as a search to find where the fault originates in the system. They are faced with a system which they suppose has been working properly, and they are searching for the location of the discrepancy between normal and defective states. They do not see the task as a more general problem-solving task in order to understand the actual functioning of the failed system and thus to explain why the system has the observed faulty response. On the other hand, a design engineer does not in his normal work in a development laboratory think in terms of standards for normal operation, but considers it his task to understand the basic functioning of the system and to test observations, made during his experiment, against his conceptual intentions.

Are the procedures found in the protocols rational? What is rational depends upon the performance criteria adopted by the person. Normally a reasonable criterion for a maintenance technician is to locate the fault as quickly as possible, and only in special circumstances will his criterion be that of minimizing the number of measurements as discussed above. From this point of view, the procedures found in our records are rational since in most cases the faults were found within very reasonable time. A good demonstration of the difference in troubleshooting time involved in the normal routine cases and those calling for more elaborate methods are given in Wohl's (1981) analysis of time distribution in field repair data. The system designer with his theoretical background may quite naturally value as rational the "elegant" deductive procedure which is informationally very efficient and based upon few observations, but this criterion is not the appropriate one on the basis of which performances in real-life maintenance work can be judged.

In the present context of supervisory control, this experiment serves to identify several diagnostic strategies which have very different requirements with respect to information and processing resources. It also illustrates the kind of subjective performance criteria which may control the choice of strategy in the actual situation.

VI. GENERALIZED STRATEGIES FOR STATE IDENTIFICATION AND DIAGNOSIS

Based on the results of this study of mental procedures in real-life diagnosis, it is now possible to generalize, and to propose a set of possible strategies, which can be used for design of operator support systems in process plant control. In this section, the typology of such a set of possible strategies will be developed (Rasmussen, 1981).

As discussed in the previous section, the object of the diagnostic search in a supervisory control task may vary: to protect the plant, search may concentrate on patterns of critical variables related to stereotypical safety actions; to compensate for the effects of the change, searching for alternative functional paths bypassing the effect of the change will be appropriate; to restore normal state, search in terms of the initiating physical change is necessary. In consequence, the diagnostic task implies a complex mental process which depends very much on details of the actual situation as well as the operator's skill and subjective preferences.

Fortunately, to support systems design, it is not necessary to have detailed process models of the mental activities which *are used* by the operators. System design can be based upon higher level structural models of the mental activities which the operators *can use* and their characteristics with respect to human limitations and preferences. If such a design is successful, operators can adapt individually and develop effective data processes.

In the system design context, a description of mental activities in information processing concepts is preferable since it is compatible with the concepts used for design of the data processing and control equipment. For this purpose, the human data processes can be described in terms of data, models and strategies. *Data* are the mental representations of information describing system state which can be represented on several levels of abstraction which in turn specify the appropriate information coding for the data presentation. *Models* are the mental representation of the system's anatomical or functional structure at the level of the search. The model can be related directly to the appropriate display formats. *Strategies* here are taken as the higher level structures of the mental processes, and they relate goals to sets of models, data and tactical process rules.

Within this framework, the actual performance of operators can be described in terms of *tactical rules* used to control the detailed processes within a formal strategy, together with the rules representing the shifts between the formal strategies which take place. During real-life performance such shifts occur frequently when difficulties are met in the current strategy or information is observed which indicates immediate results from another strategy. Since the shifts are controlled by detailed person and situation dependent aspects, they result in the very individual course and impulsive appearance of the mental processes which was discussed in the previous section.

A.   Typology of Diagnostic Search Strategies

In general, the diagnostic task implied in supervisory systems control is a search to identify a change from normal or planned system operation in terms which can refer the controller to the appropriate control actions. Such a diagnostic search can be performed in basically two different ways.

A set of observations representing the abnormal state of the system—a set of symptoms—can be used as a search template in accessing a library of symptoms related to different abnormal system conditions to find a matching set. This kind of search will be called *symptomatic search*. On the other hand, the search can be performed in the actual, maloperating system with reference to a template representing normal or planned operation. The change will then be found as a mismatch and identified by its location in the template. Consequently, this kind of search strategy was called *topographic search* in the previous section.

The difference between the two kinds of search procedures is related to a basic difference in the use of the observed information. Every observation implies identification of an information source and reading of the content of the message. By symptomatic search, reference to the identity of system state is obtained from the message read; by topographic search, reference is taken from the topographic location of the source, while the messages are subject only to good/bad judgements which are used for tactical control of the search.

B.   Topographic Search

The topographic search is performed by a good/bad mapping of the system through which the extent of the potentially "bad" field is gradually narrowed down until the location of the change is determined with sufficient resolution to allow selection of an appropriate action. The domain or level in the abstraction hierarchy in which the search is performed will vary. The search can be performed directly in the physical domain but, in most cases, the search is a mental operation at a level of abstraction which depends upon the immediate goal and intention of the controller and upon the form of the reference map or model available. Also the resolution needed for the final location depends upon the actual circumstances.

The topographic strategy is illustrated by the information flow graph of Figure 6. The main elements of the strategy which will be considered in more detail are the *model* of system used to structure the search; the kind of *data* used to represent the actual, failed plant state and the normal, reference state; and finally, the *tactical process rules* used to control the search sequence.

The topographic search is performed as a good/bad mapping of the system which results in a stepwise limitation of the field of attention within which further search is to be considered. The search depends on a map of the system which gives information on the location of sources of potential observations for which reference information is available for judgements. The map is a *model* which may identify the potential sources of

TOPOGRAPHIC SEARCH

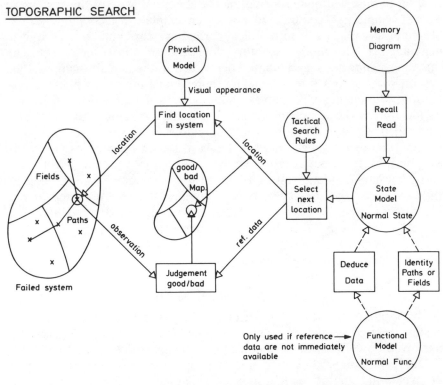

Figure 6. Information flow map illustrating the topographic search strategy, which is based on good/bad judgements of variables along a path or of patterns related to a field.

Source: Figures 6-8 are reproduced from Rasmussen, 1981, with permission from Plenum Press.

observations relative to the topology of the physical system itself, of its internal anatomical or functional structure, or of its external purposes. The search sequence is based on a set of, often heuristic, *rules* serving to limit the necessary field of attention. If different external functions can be related to separate internal parts or subsystems, a good/bad scan of external functions effectively identifies the internal field for further search. If a faulty input/output relation is found, the related causal route should be searched, e.g., by the half-split heuristic, etc. In the pure form, the tactical search decisions are based exclusively on the one bit of information obtained from the good/bad judgement of the individual observations.

The information available in observations is used rather uneconomically by topographic strategies, since they depend only upon good/bad judgements. Furthermore, they do not take into account previously experienced faults and disturbances. Therefore, switching to other strategies may be necessary to reach an acceptable resolution of the search or to acquire good tactical guidance during the search. However, the topographic search is advantageous because of its dependence upon a model of *normal* plant operation—which can be derived during design or obtained by data collection during normal operation. Therefore, consistency and correctness of the strategy can be verified and, since it does not depend on models of malfunction, it will be less disturbed by multiple or "unknown" disturbances than strategies based on disturbance symptoms.

C.   Symptomatic Search

Symptomatic search strategies are based on the information content of observations to obtain identification of system state, instead of the location of the information source in a topographic map. The search decisions are derived from the internal relationship in data sets and not from the topological structure of system properties. In principle, a search is made through a set of abnormal data sets, "symptoms," to find the set which matches the actual observed pattern of system behavior. The reference patterns can be collected empirically from incidents of system maloperation or derived by analysis or simulation of the system's response to postulated disturbances. Furthermore, reference patterns can be generated on-line if the controller has a functional model available which can be modified to match a current hypothesis about the disturbance.

When the diagnosis is performed in the data domain by a search through a library of symptom patterns, it has no logical relation to system function. The result is directly the label of the matching symptom pattern which may be in terms of cause, effect, location or appropriate control action

PATTERN RECOGNITION

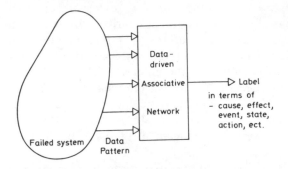

DECISION TABLE SEARCH

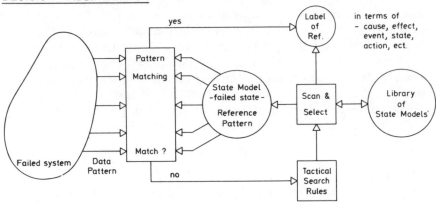

Figure 7.   Information flow maps for symptomatic diagnosis based on pattern recognition or search through a library of symptoms.

directly. Depending upon the structure of the controller and its memory, the search can be a parallel, data driven *pattern recognition*, or a sequential *decision table search* as illustrated by Figure 7.

Pattern recognition plays an important role in human diagnosis; it can efficiently identify familiar system states and disturbances directly, but it is also used frequently during topographic search to guide tactical decisions. Recognitions are then typically based on more fuzzy or general reference symptoms in terms of generic fault patterns referring to *types* of function or physical part, such as noise characteristics, instability, or forms of non-linearity.

Decision table search depends upon a set of tactical rules to guide the search which can be based on probability of occurrence, a hierarchical structuring of the attributes (like Linné's generic system for botanical identification as used in field guides), or functional relations, stored as fault trees, etc. Human diagnosticians probably would use decision tables for verification of more ambiguous recognitions.

If a search is based on reference patterns generated "on-line" by modification of a functional model in correspondence with a postulated disturbance, the strategy can be called *search by hypothesis and test* as illustrated in Figure 8. The efficiency of this type of search depends upon the tactics of generating hypotheses. Typically, in human diagnosis, hypotheses result from uncertain topographic search or fuzzy recognitions.

Symptomatic search is advantageous from the point of view of information economy, and a precise identification can frequently be obtained in a one-shot decision. One serious limitation will be that a reference pattern of the actual *abnormal* state of operation must be available, and

HYPOTHESIS & TEST

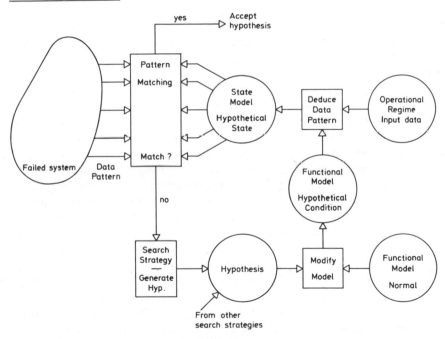

Figure 8. Information flow map for symptomatic search by hypothesis and test. The figure illustrates a conceptual test. In practice, the test may be performed by correcting the system according to the hypothesis and test for normal, instead of modifying the model as shown.

multiple faults and disturbances must be considered. This means that reference sets must be prepared by analysis or recorded from prior occurrences. Or, the reference sets can be generated on-line by means of a functional model of the system which can be modified on occasion to stimulate the abnormal system state in accordance with the current hypothesis.

D.   Resource Requirements of the Diagnostic Strategies

The information flow maps of the diagnostic strategies shown in Figures 6–8 represent the structure of strategies only, together with information about necessary data processing models. Since they do not contain information on the data-processing sequence which will be used in a specific situation, they can be used to represent a whole family of diagnostic processes typical for classes of diagnostic situations. This means that they are well suited to serve as the basis for interface design and human/computer task allocation. Task allocation implies the matching of the resource requirements from the different strategies with the resources available for the system designer, the human operator, and a process control computer for coping with the diagnostic task during different situations. In this section, the resource requirements of the different strategies will be discussed in some detail, in order to illustrate the kind of design decisions which must be made.

Consider first the *topographic search* which basically is a good/bad mapping with reference to a topographical map of the system supplying reference data for judgement. The proper level of topological representation and hence the nature of reference data and observations depend on the immediate control task. For overall control of coolant inventory and energy flow in a power plant, a representation of the mass and energy flow topology will serve the purpose of state identification in terms of deviations from specified operational state. For fault location in a system, a topological representation at the level of physical function identifying components and their individual function is necessary. With respect to resource requirements, an important feature is that the search is based on a model of the *normal* functional topology and state, which is independent of the actual situation although, as we have seen, different representations of the topology may be appropriate for different situations. However, a proper set of reference models can be prepared in advance by the systems designer, either analytically or by simulation, or it can be developed by measurement upon the operating plant. Use of the process computer for data collection and analysis can lead to a currently updated reference model of normal state, not only serving as reference during acute diagnostic situations but also for detecting gradual changes in performance.

The key process of the diagnostic search is the judgement of discrepancy between the actual and the target, i.e., the normal, state, at the relevant level of topological representation. The judgement itself is a simple process. However, preparing information for the search implies information processes having a high demand on processing and memory capacity: The proper reference model must be selected and updated from information about the defined operational regime and the actual system configuration with respect to switching and valving. Then, measured information from the system should be integrated and transformed to data which can be used for the good/bad judgements at the relevant levels of topological representations. This means that the physical variables measured in the system are only directly useful at the lowest levels of physical function. For search implied in supervisory control of process plants at, for instance, the level of mass and energy topology, all the available data from the system must be used to derive information on flows and levels in the mass/energy flow structures and balances. In a similar way, equipment which makes it possible to trace information flow, rather than signal flow through an information processing system, is useful for troubleshooting purposes. The reference for judgement can, for instance, be recorded "signatures," i.e., normal information values, along the paths. Since this preparation of information for topographical search implies well-structured processes with considerable requirements for processing and memory capacity, the task is well suited for computers.

The topographical search can be based on various search tactics which will lead to different resource requirements. The search can be based on judgement of the magnitude of state variables directly. If judgements along a flow path of information, energy, etc. are used, the search is similar to the topographic search used for electronic troubleshooting which was discussed earlier. If judgements of sets of variables related to specific functions or parts in the system are used, search is similar to the functional search in troubleshooting. Given a proper topological map and proper reference data, this search can be very efficient with human judges. In complex process systems, proper topological displays with data arranged for immediate visual comparison can be generated by computers leading to low resource requirements on the part of the human operator. However, search by direct judgement of the magnitude of variables can lead to difficulties, for instance in case of feedback loop effects, and it can be difficult to distinguish between disturbed and disturbing functions. Therefore, more elaborate search tactics should be considered, in particular when computer support is possible.

Very effective strategies can be developed, if they are based on relationships among variables which are invariant with the detailed operational state and, therefore, with the magnitude of the individual variable. A simple case of search based on state-invariant relationships has been mentioned in section V-A where a technician used an ohm-meter to find a faulty resistor. Search by relationships is more complex if a topographical search in a flow topology is based on conservation laws by checking mass or energy balances rather than flow variables. Lind (1981) has proposed a systematic search strategy based on inferences in the flow topology and has used AI (Artificial Intelligence) inspired programs like PROLOG for computer-based diagnosis. Since his approach aims at identification of changes at a high level of abstraction in terms of deviations from normal flow topology, dynamic properties of the system have not been considered. A systematic topographic search at the level of physical function and based on a reference in the form of the normal dynamic properties of equipment or functions has been proposed by Sheridan. This approach, however, is only aimed at identification of the location of the initial disturbance (Sheridan, 1981; Sheridan, et al. 1982).

In the case of diagnosis of a system which is not in operation, a very effective topographic search can be performed by forcing the system through a sequence of properly chosen operational states, *test states*, which affect various functions of the system in different ways and in properly selected combinations, and for which reference models can be prepared. Administration and evaluation of the test results depend then on logical combinatorical arguments. Computer-based support for such diagnosis in complex systems like the Apollo space system has been used (Furth, et al. 1967).

All the topographical strategies depend on search with reference to a model of *normal* function and, therefore, are well suited for identification of disturbances which were not previously experienced or which the designer has not foreseen. However, some of the more elaborate search procedures are based on evaluation of quantitative relationships in sets of variables and/or logical, combinatorical arguments for which human operators have only limited resources. Computer assisted topographical diagnosis therefore seems to be a promising area for further development.

The *symptomatic search strategies* are, in general, very information economic, but they can only be applied when a library of symptom patterns is at one's disposal. Depending upon the source of reference patterns available, the symptomatic search takes different forms with very different resource requirements.

State identification and diagnosis by data-driven *pattern recognition* represent in particular the human operator's perceptive recognition of familiar patterns, and will be the normal way to associate a situation to the related human activities. The requirement for conscious data processing is low, provided the necessary data are available in a format suited for immediate perceptual recognition. In automatic equipment, data-driven recognition is used in the traditional, hard-wired safety system which releases protective actions when a predetermined set of measured variables exceeds the trip limits.

Decision table search depends on the availability of a library of symptoms which are used as templates in a search for a match with the observed data. In the table, the symptoms are entries to information related to identified state. This information can be in terms of the root cause of a disturbance, statements of the task to pursue, or directly in terms of the procedure to follow. The decision table can be present as a kind of association matrix in the memory of a human operator or as a proper decision table in a process computer.

Whether the output from the table takes the form of a state identification, a task prescription or directly a procedure, depends on experience from previous cases of the operator or the ability of the control designer to predict the state and make decisions on the proper response. This means, the decision table will be a record of the results of prevous decisions by the operator himself or by the designer. How much of the decision sequence of Figure 1 the designer can prepare in advance depends on the invariance across occurrences of the class of situations considered, and on the ability of a designer to foresee the relevant conditions for decisions. What is not included in the designer's decisions, and hence in the decision table, must be left to the operator's decision during the actual situation. This can lead to a rather complex cooperation between designer, computer, and operator such as shown in Figures 9 and 10. Furthermore, if the designer does not have a well structured approach, the role of the operator and a computer may vary considerably with the plant situation and the question of responsibility will be very difficult to settle.

The requirements for processing and memory capacity for decision table storage and use are very high due to the need to represent a high number of special cases. Therefore, only for situations which operators meet regularly during work or at training simulators, is it possible to rely on human memory and know-how. Less frequent but well structured situations can only be covered if external support is available for information storage and retrieval. Traditionally, operators in process plants have been supported by event-related written procedures which leave the event identification to themselves. However, in some industries like nuclear power, a trend is found towards support in the form of symptom-based procedures also including the diagnostic phase. Likewise, several attempts have been made to automate state identification by means of computers which store the designer's analysis of fault patterns in decision table (Lees and Androw, 1981). However, it is difficult to evaluate the response of such systems to complex situations. For major disturbances of system operation it is generally a problem for a designer to generate proper reference

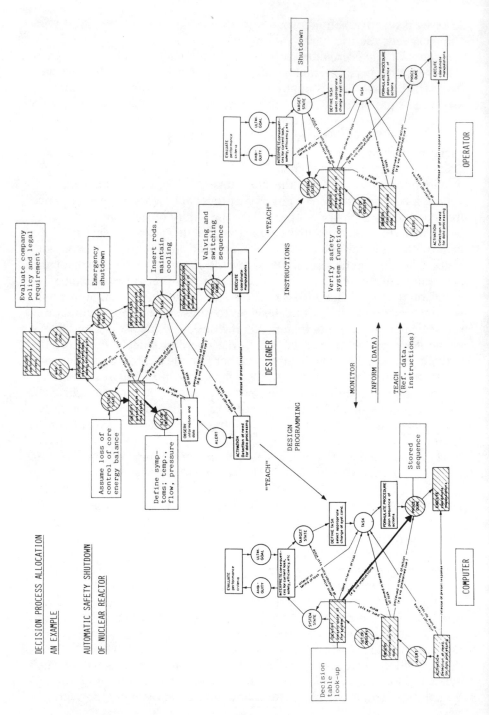

DECISION PROCESS ALLOCATION

AN EXAMPLE

AUTOMATIC SAFETY SHUTDOWN
OF NUCLEAR REACTOR

symptoms, since this requires analysis or simulation of the system outside the normal operating ranges and configurations.

A functional model for analysis or simulation of *abnormal* function is also a major requirement of *search by hypothesis and test*. A number of problems are related to the development of a functional model for this purpose. First of all, the model must represent the abnormal function of the system outside the operating ranges and the configurations which are normally considered by the models used for system design and control studies. Such models are structured as a network of quantitative, mathematical relations among physical process variables, and these relations must be known for all the relevant states of maloperation. For on-line generation of symptom patterns during a disturbance, the model can be implemented by a process computer and updated according to the actual hypothesis. An advantage would be that the model would not only be useful for hypothesis testing but also to predict, in accelerated time, responses to intended control actions. However, in addition to the problem of model range, it can be difficult to update the model to correspond to the prevailing hypothesis, since there generally will be a very complex mapping from physical changes of the system onto the corresponding changes in model structure and parameters. In particular, this is the case when model updating shall reflect a hypothesis formulated by a human operator, since this will not be in terms which are compatible with a quantitative computer model.

Functional reasoning by humans will not be based on a model in terms of relations among variables, but in complementary terms of objects and components as well as states and events as discussed in Section IV. Mapping of a hypothesis using this model is much simpler, but the symptom patterns supplied by human reasoning may be too uncertain and qualitative for a reliable test against measured data. Ultimately, human operators may, therefore, choose to test their hypothesis by correcting the systems accordingly. While this may be an effective strategy in a workshop environment, it may in process plant control lead to further complication of the situation if the hypothesis is wrong.

For man-computer co-operation in a task like search by hypothesis and test, the complementary nature of functional models suited for use by the two parts should be considered, and the techniques used in AI for computer implementation of qualitative reasoning (Williams et al., 1982;

Figure 9. The designer, the operator, and the process computer are all parts of a control decision. The figure illustrates their roles (shown hatched) in an automatic safety shut-down, in which they act "in parallel;" each with a diagnostic task based on different strategies.

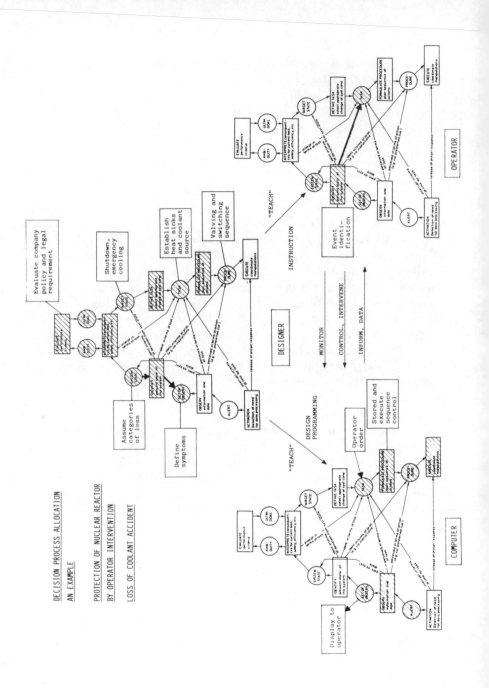

DECISION PROCESS ALLOCATION

AN EXAMPLE

PROTECTION OF NUCLEAR REACTOR
BY OPERATOR INTERVENTION

LOSS OF COOLANT ACCIDENT

Brown and deKleer, 1981) may be useful for design of intelligent interfaces.

E. Training Requirements

In general, operator training is taken care of separately from systems design by training specialists. This is most unfortunate, in particular considering future systems where interface and display formats are matched to specific tasks and situations. To do this, the designer has to consider the mental strategies which will be adopted by operators and the resources required in terms of background knowledge. Such background knowledge must include the relevant models of systems properties, which depend very much on the strategy chosen for a given task. Training can be considered a way to guide operators towards development of the mental strategies and models chosen for interface design. Therefore, design of training schemes must be integrated with the other aspects of system design.

There has recently been a trend in this direction. Shepherd et al. (1977) have studied the influence of different content of training upon diagnostic performance. They found that operators trained by the rules obtained from experienced operators were superior in diagnosis of faults not previously encountered compared with operators trained in plant theory. These in turn were superior to operators trained by practising diagnosis from symptom patterns. These differences can be readily explained, since the different training methods support the use of different strategies (i.e., topographic search, hypothesis and test, and recognition, respectively.)

Topographic search in abstract flow structures is very similar to rule-based search in "context-free" networks described by Rouse et al. (1980). It is interesting to note the observation of Rouse that the rule-based model describes the context-free strategies reasonably well but breaks down in context dependent experiments. This is probably because the context initiates shifts to symptomatic strategies which are depending upon the individual subject's prior experience. Rouse and co-workers have later (e.g., Rouse, 1983) defined expert diagnosticians as those who know when it is time to leave the symptomatic search in favor of topographical search in unfamiliar cases. An approach to training of general diagnostic skills,

---

Figure 10. The figure illustrates the roles of the designer, the operator, and a computer in a control decision which has not been fully automated, i.e. the operator and computer act "in series." The designer has automated a repertoire of protective sequences, but left the diagnosis to an operator. In addition to the decision functions, the designer, operator, and computer support each other in different "inform/teach/learn" functions (Sheridan, 1982).

PERFORMANCE FACTOR \ STRATEGY	TOPOGRAPHIC SEARCH	RECOGNITION	DECISION TABLE	HYPOTHESIS AND TEST
TIME SPENT	-	LOW	-	-
NUMBER OF OBSERVATIONS	HIGH	LOW	-	LOW
DEPENDENCY ON PATTERN PERCEPTION	-	HIGH	-	-
LOAD UPON SHORT TERM MEMORY	LOW	LOW	HIGH	HIGH
COMPLEXITY OF COGNITIVE PROCESSES	LOW	LOW	-	HIGH
COMPLEXITY OF FUNCTIONAL MODEL	LOW	-	-	HIGH
GENERAL APPLICABILITY OF TACTICAL RULES	HIGH	-	-	LOW
DEPENDENCY ON MALFUNCTION EXPERIENCE	LOW	HIGH	-	LOW
DEPENDENCY ON MALFUNCTION PRE-ANALYSIS	-	-	HIGH	-

Figure 11. The table illustrates the difference in resource requirements of the various diagnostic strategies which makes possible a resource/demand matching by proper choice of strategy. Features not filled in are either not typical, being dependent on circumstances, or not relevant.

based on training in varying problem contexts, which supports the use of topology related rather than symptom related strategies, seems to be promising (Rouse, 1982).

The conclusion here is that the design of training schemes for operators must be based on identification of the mental models and strategies which will be effective for the tasks the operators are supposed to perform. These models and strategies must then be used for integrated design of communication interface and training schemes. An illustrative review of the resource requirements of the various diagnostic strategies is given in Figure 11.

VII.   DESIGN OF SUPERVISORY CONTROL SYSTEMS

Introduction of new technology such as computer-based information processing necessitates a re-consideration of the basis for design of industrial control system. A gradual up-dating of a previous design in response to operational experience is very likely to lead to systems which are suboptimal and unnecessarily complicated. As an example, one can mention the use of computers to analyze alarm signals in process plant control rooms due to their high number—which primarily is due to the traditional one-sensor-one-indicator technology with limit sensing of measured variables individually.

A supervisory control system is basically a feedback system with the task to monitor the actual operating state of the system, and to keep it within the specified target domain. As is the case for any feedback control system, the design must be made without knowledge of the detailed nature of the disturbances the system should cope with. The design must be based on identification of the categories of disturbances and control tasks and upon consideration of the functional resources necessary to be able to meet these categories. This is in particular the case for systems with highly adaptive features, such as supervisory control systems including human operators.

This section considers how the categories and models discussed in the previous sections can be used to formulate the control requirements of the systems and identify the possible strategies which can be used as control algorithms by the supervisory control system. This kind of analysis is necessary to bridge the gap between traditional tools and methods of control system designers and recent results of research within cognitive science. Unless the control systems designer also identifies the set of information processing strategies which will meet the control requirements, he will not be able to ask meaningful questions to cognitive science concerning human capabilities and preferences. A possible way to structure such a systematic design process aiming at introduction of advanced information technology is as follows.

During the design of the physical process system, the functions of the system and their physical implementation are developed by iteratively considering the properties of the system at the various levels of the abstraction hierarchy and in increasing degree of detail as shown in Figure 12. As the degree of physical detail increases during design, so does the number of degrees of freedom in the functional states. The constraints which are necessary to maintain operation within the specified target domain must therefore be identified by the designer.

In this way, the desired states of functions and equipment will be identified during design at different levels of abstraction, and the necessary control constraints will be identified in terms of the conceptual framework related to the individual levels. In general, a skilled designer will immediately be able to identify suitable and familiar control systems concepts. However, when new technology is to be introduced for control implementation, it is necessary to develop a consistent description of the total control system function in a uniform, device independent terminology. This means that the control requirements identified at the different levels must be expressed in device independent information processes.

The system's control requirements are derived from the necessary relations between the possible system states, the specified target states, and the actions required to maintain control of the system. For a device independent formulation of the information processes involved in supervisory control, the decision sequence of Figure 1 can be used. Depending upon the control task allocation, the different phases of the decision se-

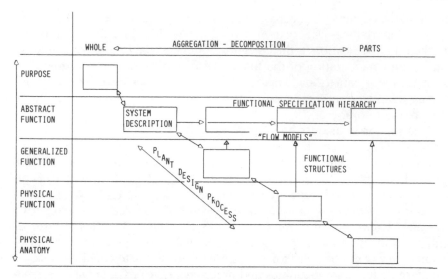

Figure 12. The design of a process plant involves an iteration simultaneously in a whole-part and an abstract-concrete dimension. As concrete details emerge, the related possible functions must be constrained by a control system. Systematic design of a supervisory control system requires that the functional specifications and the control requirements are identified in terms of device independent information processes, before implementation and man-computer task allocation are considered.

quence will be performed by the designer himself, the system operator, or the process computer.

For well-structured situations, for instance start-up and shut-down sequences, the designer can carry through all the decision steps, except the execution itself, and store the design in a decision-table in a computer or a set of operational instruction for an operator. For many other situations, the designer may be able to plan proper tasks and procedures, for instance for system protection, but not be able to foresee the related disturbed system states. In this case, diagnosis must be performed on-line by operator and computer in various degrees of cooperation. Finally, situations occur where all decision phases rely on the human operator, but since the necessary information must be supplied to the operator, his preferred strategies and related information needs must be considered when designing the interface.

In all three cases, there will be an intimate cooperation between designer, operator and computer during supervisory decision making as previously illustrated in Figures 9 and 10. Even when the computer performs stereotypical control based on design decisions, the operator will be required, or desire, to monitor performance, which means he must be supplied with information on system states and designers' intentions enabling him to verify decisions by his own preferred decision strategy.

In conclusion, design of supervisory control systems based on advanced information technology involves several distinct phases related to the concepts discussed in the previous sections and shown in Figure 13. This figure illustrates a clear, well ordered design sequence for sake of clarify.

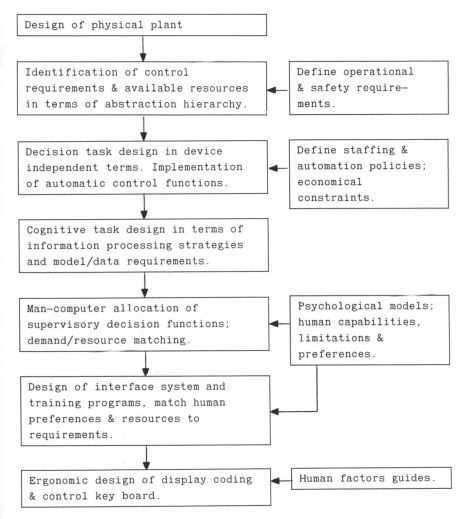

Figure 13. Phases in a systematic design of supervisory control systems.

The real design task will, however, be a complex iteration back and forth between different phases. Nevertheless, this will not affect the main lines of the present argument.

First, the *control requirements* of the system are identified in terms of target states related to different operating regimes together with the relevant categories of possible normal and disturbed states. This identification must be performed at each level of the abstraction hierarchy to develop the context of the control task, including identification of task specification (why) from the level above, and of the available resources (how) from the level below the particular task level.

Second, the *decision task* necessary to meet the control requirements is analyzed in terms of a device independent framework like the decision ladder of Figure 1. During the analysis, the conditions for the different decision phases are evaluated in the light of the manning and automation policy of the particular operating organization and the safety requirements imposed. Then, one evaluates the extent to which stereotype by-passes in the decision sequence can be analyzed and implemented as automatic functions in order to simplify the decision task during actual operations for well structured and foreseen situations.

The third state is a *cognitive task analysis and design* serving to identify and describe the possible information processing strategies which can be used for the various phases of the decision sequence. These strategies should be considered for the decision making directly involved in the supervisory control as well as for independent monitoring and support. Further, the related requirements for data and information processing models must be analyzed. For each phase of the analysis, evaluation and planning stages, several information strategies will be possible, both when involved in the control itself and when serving for monitoring and support. As it has been discussed in previous sections, these strategies will have very different resource requirements which match human operators and computer differently. Therefore, the role of operators and computers as controller and independent support and monitor is quite likely to vary during a decision task.

Therefore, the fourth step in a systematic design will be to evaluate the match between the resources available for implementation by means of human operators and computers, and to perform a *cognitive task allocation*. For this, a set of models of human information capabilities and limitations is necessary together with knowledge of the subjective task formulation and performance criteria which are likely to control the choice of strategy in the actual situation. The nature of these aspects has been illustrated for a troubleshooting task in a previous section. The need is clearly not for one comprehensive model of human operator performance, but for several different models related to different traditional fields of psychological research (Rasmussen, 1983).

This is also the case for the fifth phase, the *design of an interface system* with the information coding and display formats which will serve to match human preferences in a way that activates the proper strategies. During this phase, the design of a training scheme matching the requirements of the cognitive tasks must also be considered.

An important aspect of this approach is that only after considering the first three phases will the system designer be able to pose questions to cognitive psychology and training research. The cognitive task analysis is not a psychological problem, but a question of identification of the data processing strategies which *can* serve the control requirements of the system. These strategies must be formulated during control system design, but in system independent terms which make it possible to relate capability requirements to the resources offered by humans and computers. Only then can system designers use the models of human data processing emerging from cognitive science.

APPENDIX: USE OF VERBAL PROTOCOLS

For those who may want to use verbal protocols, the following experience from the analysis of troubleshooting in an electronic maintenance workshop may be of interest. In order to obtain a qualitative formulation of the basic features of a diagnostic procedure rather than quantitative data from standardized experiments, records were taken of several individuals locating different faults in electronic instruments. This approach enhanced the confidence of the group of technicians in that it removed the possibility of a comparison of efficiency across personnel.

In order to enable the men to familiarize themselves with the experimental procedure and the tape recorder, the instruments used for the initial cases were selected by the technicians themselves when they found it convenient to make a recording. It was determined that stereotypical procedures were used for the more simple instruments, such as amplifiers, power supplies, etc., and the cases to be recorded were then selected by the planners of the experiment.

More complex systems were then chosen (i.e., systems of a greater variety in the external display of features related to the internal operational state). These included multichannel analyzers, oscilloscopes, TV-receivers and digital voltmeters. It was found from the analysis that information available to a technician from the normal instrument displays without measurements in the circuitry was not used to any great extent. Therefore, in later phases of the experiment, simulated faults were introduced in order to have a reasonable number of cases in which the fault could be located rather precisely without opening up the instrument for measurements, if the information available from the external display was taken into careful consideration. The simulated faults were not artificial, however, but normal or probable component faults in carefully selected parts of the system. During the experiments, instruments chosen for each individual were such that he was familiar with the use of the instrument and its overall design. In order to avoid the possibility of the technician going directly to the fault by a simple recognition of the faulty response pattern, the individual technician was not given instruments for which he was an expert troubleshooter.

A total of 45 cases were recorded, comprising six individuals performing fault-finding in eight different types of instruments, each case having its particular fault. The experiment thus covered a small number of cases with a very large number of parameters influencing task performance, and reliable quantitative information was not expected. All records were preliminarily analyzed, and some of them were rejected from further analysis. Cases from the initial training of the recording were omitted, as well as cases with faults in system parts for which the group had no equipment, such as high-frequency parts of TV-receivers, and cases of intermittent or multiple faults, etc. The final, detailed analysis covered 30 recordings.

The task of finding the structure of verbal records made by several persons trying to locate individual faults in different types of systems makes one realize how colorful real-life working conditions are and how much work is needed to carry through the analysis. An almost immediate experience is the danger that the analyst himself develops fixed routines in the analysis in order to be able to manage the classification of the multitude of situations. It proves imperative to have long breaks in the analyzing effort, to be able to relax from fixations and to return to the original material with an open mind. It also seems important to have several analysts criticizing each others' models of the structure in order to break fixations and to decide which differences in the classifications made by the different analysts result from weaknesses in the definition of the classes and which are due to differences in the interpretation of the records.

The technicians were asked to relax and tell what they were thinking, feeling and doing, and to express themselves in everyday terms including short hints in fast work sequences. A record was immediately typed out and the man was asked to read it in the actual working position in front of the instrument to correct mistakes and supply supplementary information when he felt something was missing. At the same time the analyst had the first review of the record and a short talk with the man to clarify weak passages in the verbalization. The initial systematic analysis was based upon the definition of a set of elementary events describing the microstructure of the sequences. The records were coded and a computer print-out made, giving a graphical picture of the sequence as well as a connectivity matrix describing each case. The graphical read-out was found to be a convenient support in the effort to locate and identify recurrent routines. Owing to the large number of parameters, the connectivity matrices gave only a very few hints about the general pattern.

Based upon the graphic read-out, sequences identified as recurrent routines were re-analyzed from the original records and classified according to the characteristics of the data handling taking place. The analyses started by extracting the most obvious and frequent routines, leaving for later analysis the complicated and more individual parts of the records. Contributing greatly to the amount of work was the necessity for a highly iterative classification; each new class of routines which was introduced made it necessary to review all classes already used. For each subroutine a graphic symbol was chosen, with a set of codes as a graphical pattern showing the interconnection of the subroutines and with comments to facilitate later reviews.

In formulating the generalized subroutines in a subject's data processing, one has to cut away most of the details related to the individual case. To be able to find out why a specific routine is chosen, one must, however, return to the details in the original record and to the features of the specific instrument. It is important for the analysts to have a background in engineering in order to be able to imagine themselves in the task situation, and thus have a clear understanding of the meaning of the manipulations and measurements. They can then formulate *what* the technician is doing, and thus find the structure of the information handling. On the other hand, some knowledge of psychology is needed to explain *why* the man has chosen that particular approach and to formulate the goal and motivation which control the sequences found. The present discussion mainly considers the structure of technicians' information handling procedures.

During the analysis it was found that the following cases of interference from the verbalizing task may cause uncertainty during classifications of subroutines:

- It may be more attractive for the technician to do something physically, to manipulate, than to use mental activities such as reasoning because action is more readily explained in the records and goes better with the pace of speech.
- Sometimes one gets the impression that activities reported sequentially would normally be part of parallel data processing, i.e., a set

of automated routines; thus a routine may be disturbed because it is forced into consciousness, and therefore it may be erroneous and incomplete.

- Some of the records indicate that the technicians subconsciously collected information concurrently with the reported activities and that such information supports "bright ideas," which are difficult to explain later in the sequence.

However, careful discussion with the technicians after analysis of the records did not indicate any serious misinterpretation of the general structure of the procedures. When the technicians were familiar with the process of verbalization after the first few cases, they found the execution of the task as well as the time spent in locating the faults to be normal.

REFERENCES

Alexander, C. *Notes on the Synthesis of Form*. Cambridge, MA: Harvard Press, 1964.

Bartlett, F. *Thinking, An Experimental and Social Study*. London: Unwin, 1958.

Brown, I. S. and deKleer, J. Towards a theory of qualitative reasoning. In J. Rasmussen and W. B. Rouse (Eds.), *Human Detection and Diagnosis of System Failures*. New York: Plenum Press, 1981, pp. 317–335.

Bruner, J., Goodnow, J. J. and Austin, G. A. *A Study of Thinking*. New York: Wiley, 1956.

Cassierer, E. *Substance and Function*. Chicago: Open Court Publishing Co., Dover Edition; New York: Dover Publ., 1953 (1923).

Dennett, D. C. Intentional systems. *Journal of Philosophy LXVIII* (4), 1971.

Eastman, C. M. The representation of design problems and maintenance of their structure. In Latombe (Ed.), *Artificial Intelligence and Pattern Recognition in Computer-Aided Design*. New York: North-Holland, 1978.

Furth, E., Grant, G., and Smithline, H. Data conditioning and display for Apollo prelaunch checkout. Dunlap and Associates, NASA, N-68-12531, 1967.

Lees, F. P. Computer support for diagnostic tasks in the process industries. In J. Rasmussen and W. B. Rouse (Eds.), *Human Detection and Diagnosis of System Failures*. New York: Plenum Press, 1981, pp. 369–388.

Lind, M. The use of flow models for automated plant diagnosis. In J. Rasmussen and W. B. Rouse (Eds.), *Human Detection and Diagnosis of System Failures*. New York: Plenum Press, 1981, pp. 411–432.

Lind, M. Multilevel flow modelling of process plant for diagnosis and control. *International Meeting on Thermal Nuclear Reactor Safety*, Chicago, 1982. (Also: Roskilde, Denmark: Riso National Laboratory, Report No. M-2357.)

Polanyi, M. *Personal Knowledge*. London: Routledge & Kegan Paul, 1958.

Polayni, M. *The Tacit Dimension*. New York: Doubleday, 1967.

Rasmussen, J. Outlines of a hybrid model of the human process operator. In T. B. Sheridan and G. Johannsen (Eds.), *Monitoring Behavior and Supervisory Control*. New York: Plenum Press, 1976, pp. 371–382.

Rasmussen, J. On the structure of knowledge—A morphology of mental models in a man-machine context. Roskilde, Denmark: Riso National Laboratory, Report No. M-2192, 1979.

Rasmussen, J. Models of mental strategies in process plant diagnosis. In J. Rasmussen and W. B. Rouse (Eds.), *Human Detection and Diagnosis of System Failures*. New York: Plenum Press, 1981, pp. 241–258.

Rasmussen, J. Skills, rules, knowledge; signals, signs, and symbols; and other distinctions in human performance models. *IEEE Transactions on Systems, Man, and Cybernetics SMC-13*(3), 1983.

Rasmussen, J. and Jensen, A. A study of mental procedures in electronic troubleshooting. Roskilde, Denmark: Riso National Laboratory, Report No. M-1582, 1973.

Rasmussen, J. and Jensen, A. Mental procedures in real life tasks: A case study of electronic trouble shooting. *Ergonomics 17*(3):293–307, 1974.

Rasmussen, J. and Lind, M. Coping with compexity. *European Annual Conference on Human Decision Making and Manual Control*, Delft, 1981. (Also in: Roskilde, Denmark: Riso National Laboratory, Report. No. M-2293.)

Rouse, W. B. A mixed-fidelity approach to technical training. *Journal of Educational Technology Systems 11*(2):103–115, 1982.

Rouse, W. B. Models of human problem solving: Detection, diagnosis, and compensation for system failures. *Automatica 19*, 1983.

Rouse, W. B., Rouse, S. H. and Pellegrino, S. J. A rule-based model of human problem solving performance in fault diagnosis tasks. *IEEE Transactions on Systems, Man, and Cybernetics SMC-10*(7):366–376, 1980.

Russell, B. On the notion of cause. *Proc. Aristotelean Society 13*:1–25, 1913.

Shepherd, A., Marshall, E. C., and Duncan, K. D. Diagnosis of plant failures from a control panel: A comparison of three training methods. *Ergonomics 20*:347–361, 1977.

Sheridan, T. B. Understanding human error and aiding human diagnostic behavior in nuclear power plants. In J. Rasmussen and W. B. Rouse (Eds.), *Human Detection and Diagnosis of System Failures*. New York: Plenum Press, 1981, pp. 19–35.

Sheridan, T. B., Tsach, U. and Tzelgow, J. A new method for failure detection and location in complex dynamic systems. *Proc. of American Control Conference*, Arlington, Virginia, June 14–16, 1982.

Williams, M. D., Moran, T. P., and Brown, J. S. The role of conceptual models in nuclear power plant operations. *Proc. of Workshop on Cognitive Modeling of Nuclear Plant Control Room Operators*, Nuclear Regulatory Commission, Dedham, Massachusetts, August, 1982.

Wohl, J. System complexity, diagnostic behavior and repair time. In J. Rasmussen and W. B. Rouse (Eds.), *Human Detection and Diagnosis of System Failures*. New York: Plenum Press, 1981, pp. 217–231.

HUMAN PROBLEM SOLVING IN FAULT DIAGNOSIS TASKS

William B. Rouse and Ruston M. Hunt

ABSTRACT

A series of experimental studies and mathematical models of human problem solving are summarized. Emphasis is on the differences between general
and specific problem solving abilities in fault diagnosis tasks. Results of
evaluating a variety of methods for training and aiding problem solvers are
discussed. Conclusions are drawn regarding the role of humans in dealing
with system failure situations.

I. INTRODUCTION

One of the reasons often given for employing humans in systems is their
supposed abilities to react appropriately and flexibly in failure situations.
On the other hand, one seems to hear increasingly about failure situations
being aggravated by "human error." The apparent inconsistency of these
two observations can cause one to wonder what role the human should
actually play (Rasmussen and Rouse, 1981).

Advances in Man-Machine Systems Research, Volume 1, pages 195–222.
Copyright © 1984 by JAI Press Inc.
All rights of reproduction in any form reserved.
ISBN: 0-89232-404-X

This question has led the authors and their colleagues to the pursuit of a series of investigations of human problem solving performance in fault diagnosis tasks. Using three different fault diagnosis scenarios, several hundred subjects (mostly maintenance trainees) have been studied in the process of solving many thousands of problems. The results of these studies have motivated the development of several mathematical models of human problem solving behavior. The three tasks, results of ten experiments, and five models are reviewed in this chapter.

Besides trying to assess problem solving abilities, considerable effort has also been invested in studying alternative methods of training humans to perform fault diagnosis tasks. One issue that has been particularly intriguing concerns the extent to which humans can be trained to have general, context-free problem solving skills. From a theoretical point of view, it is of fundamental interest to know whether skills are context-free or context-specific. From a practical perspective, this issue is perhaps even more important in terms of training personnel to serve in multiple domains (e.g., to diagnose faults in a wide variety of systems). This chapter considers the extent to which the studies discussed here have provided an answer to the context-free versus context-specific question.

The overall goal of this research has been to determine an appropriate role for humans in failure situations and, to develop methods of training humans to fill that role. In a final section of this chapter, the variety of results presented here will be used as a basis for proposing how these issues should be resolved.

II. FAULT DIAGNOSIS TASKS

Three types of fault diagnosis task were used in this research. Two types involve computer simulations of network representations of systems in which subjects are required to find faulty components. The third type involves troubleshooting of real equipment. The three types of task represent a progression from a fairly abstract simulation that includes only one or two basic operations, to a somewhat realistic simulation and, finally, to real equipment.

A. TASK

In considering alternative fault diagnosis tasks for initial studies, one particular task feature seemed to be especially important. This feature is best explained with an example. When trying to determine why component, assembly, or subsystem A is producing unacceptable outputs, one may note that acceptable performance of A requires that components B, C, and D be performing acceptably since component A depends upon them. Further, B may depend on E, F, G, and H while C may depend on F and G, and so on. Fault diagnosis in situations such as this example involves dealing with a network of dependencies among components in terms of their abilities to produce acceptable outputs. The class of tasks described in this paragraph was the basis for the task chosen for initial investigations. Because this type of task emphasizes the structural properties of systems (i.e., relationships among components), the acronym chosen was TASK which stands for Troubleshooting by Application of Structural Knowledge.

TASK involves fault diagnosis of graphically displayed networks. An example of TASK 1 is shown in Figure 1. These networks operate as follows. Each component has a random number of inputs. Similarly, a random number of outputs emanate from each component. Components are devices that produce either a 1 or 0. An output of 1 denotes an acceptable output; 0 an unacceptable output. All outputs emanating from a component carry the value produced by that component.

A component will produce a 1 if: 1) *All* inputs to the component carry values of 1 and, 2) The component has not failed. If either of these two conditions are not satisfied, the components will produce a 0. Thus, components are like AND gates. If a component fails, it will produce values

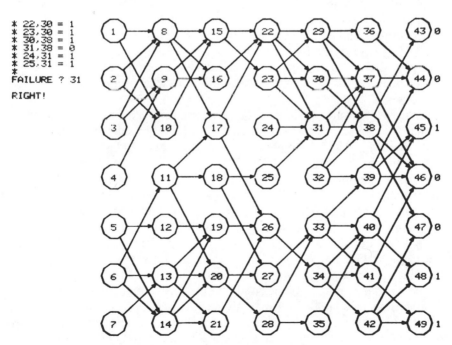

Figure 1. An Example of TASK 1.

of 0 on all the outputs emanating from it. Any components that are reached by these outputs will in turn produce values of 0. This process continues and the effects of a failure are thereby propagated throughout the network.

A problem begins with the display of a network with the outputs indicated, as shown on the righthand side of Figure 1. Based on this evidence, the subject's task is to "test" connections between components until the failed component is found. The upper lefthand side of Figure 1 illustrates the manner in which connections are tested. An * is displayed to indicate that subjects can choose a connection to test. They enter commands of the form "component 1, component 2" and are the shown the value carried by the connection. If they respond to the * with a simple "return," they are asked to designate the failed component. Then, they are given feedback about the correctness of their choice.[1] And then, the next randomly-generated problem (i.e., totally new) is displayed.

In the experiments conducted using TASK 1, computer aiding was one of the experimental variables. The aiding algorithm is discussed in detail elsewhere (Rouse, 1978a). Succinctly, the computer aid is a somewhat sophisticated bookkeeper that uses the structure of the network (i.e., its topology) and known outputs to eliminate components that cannot possibly be the fault (i.e., by crossing them off). Also, it iteratively uses the results of tests (chosen by the subject) to further eliminate components from future consideration by crossing them off. In this way, the "active" network iteratively becomes smaller and smaller.

TASK 1 is fairly limited in that only one type of component is considered. Further, all connections are feed-forward and thus, there are no feedback loops. To overcome these limitations, a second version of TASK was devised.

Figure 2 illustrates the type of task of interest. This task is somewhat similar to TASK 1 in terms of using an acceptable/unacceptable dichotomy, requiring similar commands from subjects, and so on. Only the differences between TASK 1 and TASK 2 are explained here.

A square component will produce a 1 if: 1) *All* inputs to the component carry values of 1 and, 2) The component has not failed. Thus, square components are like AND gates. A hexagonal component will produce 1 if: 1) *Any* input to the component carries a value of 1, and 2) The component has not failed. Thus, hexagonal components are like OR gates. For both AND and OR components, if either of the two conditions is not satisfied, the component will produce a 0.

The overall problem is generated by randomly connecting components. Connections to components with higher numbers (i.e., feed-forward) are equally likely with a total probability of p. Similarly, connections to components with lower numbers (i.e., feedback) are equally likely with a total probability of 1-p. The ratio p/(1-p), which is an index of the level of

```
*  20  25  =  1
*  13  24  =  0
*  15  13  =  0
*   8  15  =  0
*   1  25  =  0
*
FAILURE ?   1

RIGHT!
```

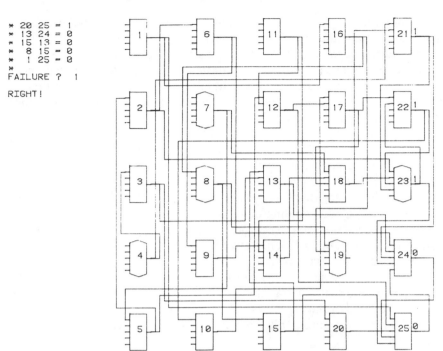

Figure 2. An Example of TASK 2.

feedback, was one of the independent variables in the experiments to be discussed later. OR components are randomly placed. The effect of the ratio of the number of OR to AND components was also an independent variable in the experiments.

B. FAULT

TASK 1 and TASK 2 are context-free fault diagnosis tasks in that they have no association with a particular system or piece of equipment. Further, subjects never see the same problem twice. Thus, they cannot develop skills particular to one problem. Therefore, one must conclude that any skills that subjects develop have to be general, context-free skills.

However, real-life tasks are not context-free. And thus, one would like to know if context-free skills are of any use in context-specific tasks. In considering this issue, one might first ask: Why not train the human for the task he is to perform? This approach is probably acceptable if the human will in fact only perform the task for which he is trained. However, with technology changing rapidly, an individual is quite likely to encounter many different fault diagnosis situations during his career. If one adopts

the context-specific approach to training, then the human has to be substantially retrained every time he changes situations.

An alternative approach is to train humans to have general skills which
they can transfer to a variety of situations. Of course, they still will have
to learn the particulars of each new situation, but they will not do this by
rote. Instead, they will use this context-specific information to augment
their general fault diagnosis abilities.

The question of interest, then, is whether or not one can train subjects
to have general skills that are in fact transferrable to context-specific
tasks. With the goal of answering this question in mind, another fault
diagnosis task was designed (Hunt, 1979; Hunt and Rouse, 1981). The
acronym chosen for this task was FAULT which stands for Framework
for Aiding the Understanding of Logical Troubleshooting.

Since FAULT is context-specific, one can employ hardcopy schematics
rather than generating random networks online such as used with TASK.
A typical schematic is shown in Figure 3. The subject interacts with this
system using the display shown in Figure 4. The software for generating
this display is rather general and particular systems of interest are completely specified by data files, rather than by changes in the software itself.
Thus far, various automobile, aircraft, and marine systems have been
simulated.

FAULT operates as follows. At the start of each problem, subjects are
given rather general symptoms (e.g., will not light off). They can then
gather information by checking gauges, asking for definitions of the functions of specific components, making observations (e.g., continuity
checks), or by removing components from the system for bench tests.
They also can replace components in an effort to make the system operational again.

Associated with each component are costs for observations, bench
tests, and replacements as well as the a priori probability of failure. Subjects obtain this data by requesting information about specific components. The time to perform observations and tests are converted to dollars
and combined with replacement costs to yield a single performance measure of cost. Subjects are instructed to find failures so as to minimize cost.

As with TASK, computer aiding was an independent variable in one
of the experiments with FAULT (Hunt, 1981; Hunt and Rouse, 1984b).
The aiding scheme monitors subjects for inferential errors (i.e., seeking
information that, by structural inference, is already available) and provides context-specific feedback concerning how the appropriate inference
could be made. Aided subjects were also allowed to test the validity of
hypotheses by asking the computer whether or not a particular component
was in the feasible set of possible failures given the information collected
up to that point.

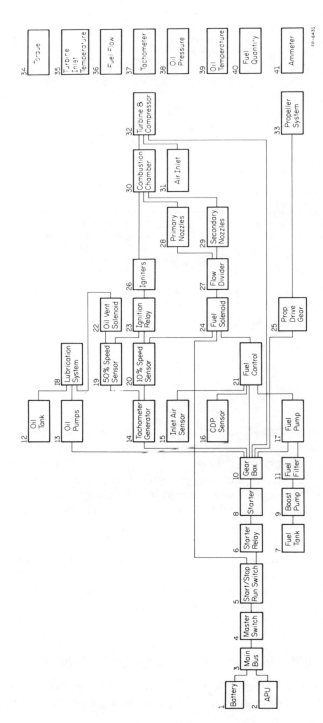

Figure 3. An Example FAULT System.

201

System: Turboprop	Symptom: Will not light off	

You have six choices:	34	Torque	
1 Observation ....OX,Y	35	Turbine Inlet Temp	Low
2 Information...........IX	36	Fuel Flow	Low
3 Replace a part.........RX	37	Tachometer	Low
4 Gauge reading.........GX	38	Oil Pressure	Normal
5 Bench test.............BX	39	Oil Temperature	Normal
6 Comparison ....CX,Y,Z	40	Fuel Quantity	
(X,Y and Z are part numbers)	41	Ammeter	Normal

Your choice ...

Actions	Costs	Actions	Costs	Parts Replaced	Costs
4, 5 Normal	$ 1			14 Tach Generator	$ 199
26,30 Abnormal	$ 1				
14,20 Not aval	$ 0				
14 is Abnormal	$ 27				

FP-6441

Figure 4.   The FAULT Display.

C.   Real Equipment

The experiments involving real equipment required subjects to diagnose failures in four and six cylinder engines typical of those used in modern general aviation aircraft (Johnson, 1980; Johnson and Rouse, 1982b). The five problems chosen for study represented four engine subsystems: electrical, ignition, lubrication, and fuel. More specifically, the five problems studied were: 1) an open starter lead, 2) a defective spark plug wire, 3) an obstructed oil fitting, 4) a defective spark plug, and 5) an obstructed fuel line.

Subjects were required to observe malfunctioning (but operating) engines and, by appropriately choosing tests, identify the source of the problem. They were supplied with all of the tools and test equipment necessary to diagnose any fault that they might encounter. Technical manuals and related information were also available.

III.   MEASURES OF PERFORMANCE

In the series of experiments to be discussed in the next section, the subjects' instructions varied as the series progressed. While the initial experiment emphasized minimizing the number of tests to diagnose the failure correctly, later experiments stressed minimum time and cost. All three of these measures reflect the *product* of fault diagnosis. While such measures may appropriately gauge the overall goals of fault diagnosis, product measures do not provide much insight into the *process* of fault diagnosis (Duncan and Gray, 1975; Brooke, et al., 1980). Much finer-grained process measures are needed to provide the desired insights into human behavior. In this section, the way in which this issue was addressed is reviewed.

A. Dimensions of Performance

Analyses of the results of the initial experiments with TASK were limited to product measures, typically adjusted for problem difficulty by normalizing with respect to optimal performance. These measures appeared to be satisfactory until experiments with FAULT were conducted. It was then found that the product measures were much too sensitive to individual differences among problems and subjects.

This realization led to the development of a variety of fine-grained process measures (Hunt and Rouse, 1981). One pair of these measures considers diagnostic costs greater than optimal (the minimum) and partitions these suboptimal costs into two categories: errors and inefficiency. Errors are defined as actions that do not reduce the size of the feasible set of failures (i.e., non-productive actions). Inefficient actions are productive but not as productive as possible. Another fine-grained measure is the expected (as opposed to actual) information gain (in bits) per action. A third measure reflects subjects' allocation of expenditures among the types of action available.

The usefulness of these process measures motivated a comprehensive investigation of performance measures (Henneman, 1981; Henneman and Rouse, 1984). A thorough review of the literature, as well as consideration of previous experience with TASK and FAULT, produced a set of twenty candidate measures. These measures were evaluated using data from two of the later experiments. Correlation, regression, and factor analyses were employed.

The results were unequivocal. Among the twenty measures, there are only three unique dimensions: errors, inefficiency, and time. Thus, a single product measure such as time or cost does not adequately describe human performance. This result also showed that the choice of process measures of errors and inefficiency, as well as the product measure of time, for the earlier studies with FAULT was very appropriate.

The emergence of errors as one of the primary dimensions of diagnostic performance led to three studies of human error and the development of a general methodology for analysis and classification of human error (van Eekhout and Rouse, 1981; Johnson and Rouse, 1982a; Rouse and Rouse, 1983). An early version of this methodology was used to analyze the results of the first real equipment experiment and produced changes in the training methods that were subsequently shown to reduce substantially the frequencies of certain types of human errors. These results are reviewed in a later section.

B.   Predictive Measures

It is interesting to consider the extent to which problem solving performance is correlated with a priori characteristics of subjects rather than the effects of training. To explore this issue, the performance of subjects on TASK, FAULT, and real equipment was correlated with twelve measures of ability, aptitude, and cognitive style (Henneman, 1981; Henneman and Rouse, 1984). Results of standard scholastic aptitude tests were used as ability measures. A mechanical reasoning test was employed to obtain a measure of aptitude. Two dimensions of cognitive style were considered: impulsivity-reflectivity (via Matching Familiar Figures Test [Kagan, 1965]) and field dependence-independence (via Embedded Figures Test [Witkin, et. al., 1971]).

Results indicated that cognitive style was a much better predictor (Pearson's r ~ 0.5) of problem solving performance than were the measures of ability and aptitude. It should be noted, however, that the trainees whose data was employed in this analysis had to meet certain standards of ability and aptitude (but not style) in order to be accepted into the training program which was studied. Thus, the fairest conclusion seems to be that cognitive style becomes dominant once minimum standards of ability and aptitude are met.

Detailed statistical analyses of the cognitive style results were performed by partitioning trainees into impulsive and reflective groups, as well as field dependent and independent groups, and using analyses of variance with dependent measures of errors, inefficiency, and time (Rouse and Rouse, 1982). The strongest conclusion to result from this analysis was that impulsives made significantly more errors. Several interesting comparisons with results published in the cognitive style literature were also found.

A further analysis of performance changes over time indicated that reflective field independents were the best problem solvers, although the superiority of field independents over field dependents tended to decrease as experience was gained (Hunt, et al., 1981). One can conjecture that the pattern recognition abilities of field dependents required more time to adapt to new problem domains; however, they did eventually adapt.

On the other hand, the effects of impulsivity were not compensated for with practice.

IV. EXPERIMENTS

Using the three tasks and variety of performance measures described in the last two sections, ten experiments were performed involving over 300 subjects who solved over 24,000 fault diagnosis problems. Over 90% of the subjects were trainees in an FAA certificate program in aircraft powerplant maintenance. The remainder were students or former students in engineering. In this section, the statistically significant results of these experiments will be reviewed.

Experiments one through five focused on problem solving performance with TASK. Experiments six through eight considered the relationships between TASK and FAULT performance. Experiments nine and ten studied transfer of training from TASK and/or FAULT to real equipment.

A. Experiment One

The first experiment utilized TASK 1 and considered the effects of problem size, computer aiding, and training. Problem size was varied to include networks with 9, 25, and 49 components. Computer aiding was considered both in terms of its direct effect on task performance and in terms of its effect as a training device (Rouse, 1978a).

Eight subjects participated in this experiment. The experiment was self-paced. Subjects were instructed to find the fault in the minimum number of tests while also not using an excessive amount of time and avoiding all mistakes. A transfer of training design was used where one-half of the subjects were trained with computer aiding and then transitioned to the unaided task, while the other one-half of the subjects were trained without computer aiding and then transitioned to the aided task.

Results indicated that human performance, in terms of average number of tests until correct solution, deviated from optimality as problem size increased. However, subjects performed much better than a "brute force" strategy which simply traces back from an arbitrarily selected 0 output. This result can be interpreted as meaning that subjects used the topology of the network (i.e., structural knowledge) to a great extent as well as knowledge of network outputs (i.e., state knowledge).

Considering the effects of computer aiding, it was found that aiding always produced a lower average number of tests. However, this effect was not statistically significant. Computer aiding did produce a statistically significant effect in terms of a positive transfer of training from aided to unaided displays for percent correct. Specifically, percent correct was greater with aided displays (98% vs. 89%) and subjects who transferred aided-to-unaided were able to maintain the level of performance achieved with aiding.

B. Experiment Two

This experiment utilized TASK 1 and was designed to study the effects of forced-pacing (Rouse, 1978a). Since many of the interesting results of the first experiment were most pronounced for large problems (i.e., those with 49 components), the second experiment considered only these large problems. Replacing problem size as an independent variable was time allowed per problem, which was varied to include values of 30, 60, and 90 seconds. The choice of these values was motivated by the results of the first experiment which indicated that it would be difficult to solve problems in less than 30 seconds consistently while it would be relatively easy to solve problems in less than 90 seconds.

This variable was integrated into the experimental scenario by adding a clock to the display. Subjects were allowed one revolution of the clock in which to solve the problem. The circumference of the clock was randomly chosen from the three values noted above. If subjects had not solved the problem by the end of the allowed time period, the display was erased and they were asked to designate the failed component.

As in the first experiment, computer aiding and training were also independent variables. Twelve subjects participated in this experiment. Their instructions were to solve the problems within the time constraints while avoiding all mistakes.

Results of this experiment indicated that the time allowed per problem and computer aiding had significant effects on human performance. A particularly interesting result was that forced-paced subjects utilized strategies requiring many more tests than necessary (i.e., greater than self-paced subjects by a factor of approximately four). It appears that one of the effects of forced-pacing was that subjects chose to employ less structural information in their solution strategies, as compared to self-paced subjects. While computer aiding resulted in significantly fewer tests (0.99 vs. 3.33) and a greater percent correct (89% vs. 80%), there was no positive (or negative) transfer of training for forced-paced subjects, indicating that subjects may have to be allowed to reflect on what computer aiding is doing for them if they are to gain transferrable skills. In other words, time pressure can prevent subjects from studying the task sufficiently to gain skills via computer aiding.

C. Experiment Three

Experiments one and two utilized students or former students in engineering as subjects. To determine if the results obtained were specific to that population, a third experiment investigated the fault diagnosis abilities of forty trainees in the fourth semester of a two-year FAA certificate program in aircraft powerplant maintenance (Rouse, 1979a).

The design of this experiment was similar to that of the first experiment in that TASK 1 was utilized and problem size, computer aiding, and training were the independent variables. However, only transfer in the aided-to-unaided direction was considered. Further, subjects' instructions differed somewhat in that they were told to find the failure in the least amount of time possible, while avoiding all mistakes and not making an excessive number of tests.

As in the first experiment, performance in terms of average number of tests until correct solution deviated from optimality as problem size increased. Computer aiding significantly decreased this deviation (0.60 vs. 1.71, or 65% better). Considering transfer of training, it was found that aided subjects utilized fewer tests to solve problems without computer aiding, particularly for the larger problems (1.11 vs. 2.12 tests greater than optimal). A very specific explanation of this phenomenon will be offered in a later discussion.

D. Experiment Four

Experiment four considered subjects' performance in TASK 2 (Rouse, 1979b). Since the main purpose of this experiment was to investigate the suitability of a model of human decision making in fault diagnosis tasks that include feedback and redundancy, only four highly-trained subjects were used. The two independent variables included the aforementioned level of feedback (i.e., $p/(1 - p)$) and the ratio of number of OR to AND components in a network of twenty-five components.

The results of this experiment indicated that increased redundancy (i.e., more OR components) significantly decreased the average number of tests (3.47 vs. 4.91) and average time until correct solution (63.3 sec vs. 101.7 sec) of fault diagnosis problems. While there were visible trends in performance as a function of the level of feedback, this effect was not significant. The reason for this lack of significance was quite clear. Two subjects developed a strategy that carefully considered feedback while the other two subjects developed a strategy that discounted the effects of feedback. Thus, the average across all subjects was insensitive to feedback levels. One of the models to be described later yields a fairly succinct explanation of this result.

E. Experiment Five

The purpose of this experiment was to investigate the performance of maintenance trainees in TASK 2, while also trying to replicate the results

of experiment three. Forty-eight trainees in the first semester of the previously noted FAA certificate program served as subjects (Rouse, 1979c).

The design involved a concatenation of experiments three and four.
Thus, the experiment included two sessions. The first session was primarily for training subjects to perform the simpler TASK 1. Further, the
results of the first session, when compared with the results of experiment
three, allowed a direct comparison between first and fourth semester
trainees.

The second session involved a between-subjects factorial design in
which level of feedback and proportion of OR components were the independent variables. Further, training on TASK 1 (i.e., unaided or aided)
was also an independent variable. Thus, the results of this experiment
allowed assessment of transfer of training between two somewhat different tasks.

As in the previous experiments, TASK 1 performance in terms of average number of tests until correct solution deviated from optimality as
problem size increased and, the deviation was substantially reduced with
computer aiding (0.57 vs. 1.53, or 63% better). Computer aiding also resulted in faster solutions (46.5 sec vs. 62.1 sec). However, unlike the
results from experiment three, there was no positive (or negative) transfer
of training from the aided displays. This result as well as subjects' comments led to the conjecture that the first semester students perhaps differed from the fourth semester students in terms of intellectual maturity
(i.e., the ability to ask why computer aiding was helping them rather than
simply accepting the aid as a means of making the task easy).

On the other hand, TASK 2 provided some very interesting transfer of
training results. In terms of average time until correct solution, subjects
who received aiding during TASK 1 training were initially significantly
slower in performing TASK 2. However, they eventually far surpassed
those subjects who received unaided TASK 1 training. This initial negative transfer (13% slower) and then positive transfer (20% faster) is an
interesting but puzzling phenomenon.

F.   Experiment Six

This experiment considered subjects' abilities to transfer skills developed in the context-free TASK 1 and TASK 2 to the context-specific
FAULT. Thirty-nine trainees in the fourth semester of the two-year FAA
certificate program served as subjects (Hunt, 1979; Hunt and Rouse,
1981).

The design of this experiment was very similar to previous experiments
except the transfer trials involved FAULT rather than the context-free
tasks. The FAULT scenarios used included an automobile engine and
two aircraft powerplants, one of which was unfamiliar to trainees. Both

TASK 1 and TASK 2 were used for the training trials. Overall, subjects participated in six sessions of ninety minutes in length over a period of six weeks.

As noted earlier, since initial analyses of the results indicated a very substantial degree of inter-subject and inter-problem variability, it was decided to employ more fine-grained measures for FAULT. One of these fine-grained measures involved partitioning subjects' suboptimality (i.e., expenditures greater than optimal) into those due to errors and those due to inefficiency. Another measure was the expected information gain (in bits) per action. A third measure reflected the subjects' allocation of expenditures among observations, bench tests, and unnecessary replacements.

Use of these fine-grained performance measures led to quite clear conclusions. Trainees who had received aided training with TASK 1 were consistently able to achieve significantly better performance on the aircraft powerplant problems ($513 vs. $578 for cost due to inefficiency), especially for problems involving less familiar powerplants. It was found that their suboptimality in terms of inefficiency could be attributed to their focusing on high cost, low information gain actions (i.e., bench tests and replacements) to a much greater extent than the optimal solution.

G. Experiment Seven

The purpose of this experiment was to replicate experiment six using first semester rather than fourth semester maintenance trainees. Sixty trainees participated. The design of the experiment was very similar to experiment six except that only TASK 1 training was used. Further, one of the aircraft powerplant scenarios was changed to allow inclusion of a more sophisticated system (Hunt and Rouse, 1981).

The results for the first semester trainees were mixed with a substantial positive transfer of aided training in terms of inefficiency ($469 vs. $1266) and a slight negative transfer of training in terms of expected information gain (0.51 vs. 0.53 bits/action). However, as with the fourth semester trainees, inefficiency could be attributed to inappropriate choices of high cost, low information gain actions.

H. Experiment Eight

This experiment considered the effects of computer-aided training with FAULT. Thirty-four first semester maintenance trainees participated in ten problem solving sessions over a ten week period. Half of the subjects received aiding while the other half did not. The two groups were initially matched on the basis of TASK 1 performance. Problems on FAULT included six different automobile and aircraft systems, some of which were unfamiliar to subjects (Hunt, 1981; Hunt and Rouse, 1984b).

The results of this experiment indicated that aiding decreased suboptimality in terms of inefficient actions for both the familiar (4.20 vs. 4.73) and unfamiliar (4.47 vs. 4.90) systems. Aiding significantly reduced the frequency of errors for the unfamiliar systems (0.40 vs. 0.83). (It is important to note that the aiding was designed to reduce errors; benefits in terms of decreased inefficiency were only a by-product of aiding.) Considering transfer from FAULT to TASK, subjects trained with aided FAULT had a lower frequency of errors with TASK (0.08 vs. 0.20).

I. Experiment Nine

The purpose of this experiment was to evaluate the transfer of training with TASK 1, TASK 2, and FAULT to real equipment (Johnson, 1980; Johnson and Rouse, 1982b). Thirty-six fourth semester trainees participated as subjects. Each subject was allocated to one of the three training groups. Groups were balanced with respect to various a priori measures (e.g., grade point average). One group was trained using a sequence of TASK 1 and TASK 2 problems. Another group was trained with FAULT. The third group, the control group, received "traditional" instruction including reading assignments, video taped lectures, and quizzes. The transfer task involved the aforementioned five problems on two real aircraft engines.

Performance measures for the real equipment problems included an average performance index based on a fine-grained analysis of each action, overall adjusted cost (based on the manufacturer's flat-rate manual), and an overall rating by an observer. Results indicated that traditional instruction was only superior if explicit demonstrations were provided for the exact failures to be encountered (i.e., three of the five real equipment problems). Otherwise, there were no significant differences among the three training methods.

More specifically, for the average performance index, which ranged from 1.0 to 5.0, the three problems which were explicitly demonstrated yielded 4.4 for traditional instruction and 3.8 for TASK and FAULT; the two problems that were not explicitly demonstrated yielded the non-significant difference of 4.4 for traditional instruction and 4.2 for TASK and FAULT. Thus, training with the computer simulations was as useful as traditional training as long as the latter form of instruction was general in nature (i.e., did not provide "cookbook" solutions for particular problems).

J. Experiment Ten

This experiment also considered transfer to real equipment, and compared a combination of TASK and FAULT to traditional instruction.

Twenty-six fourth semester maintenance trainees served as subjects. One half of the subjects were trained with TASK/FAULT where FAULT was somewhat modified to include information on how tests are physically made and how results should be interpreted. The other half of the subjects received traditional instruction similar to that in experiment nine (Johnson and Rouse, 1982b).

Based on the same performance measures as used for experiment nine, it was found that the TASK/FAULT combination was equivalent to traditional instruction for all five problems, even those for which explicit solution sequences had been provided within the traditional instruction. More specifically, the average performance index was 4.2 for traditional instruction and 3.9 for TASK/FAULT, a difference which was not statistically significant. Thus, somewhat generalized training was found to be competitive with problem-specific training. The full implications of this result will be discussed in a later section.

V. MODELS OF HUMAN PROBLEM SOLVING

The numerous empirical results of the experimental studies discussed above are quite interesting and offer valuable insights into human fault diagnosis abilities. However, it would be more useful if one could succinctly generalize the results in terms of theories or models of human problem solving performance in fault diagnosis tasks.[2] Such models will eventually be useful for predicting human performance in fault diagnosis tasks and, perhaps for evaluating alternative aiding schemes and training methods. More immediately, however, the models discussed here were of use for interpreting research results and defining the directions of the investigations.

A. Models of Complexity

It is interesting to consider why some fault diagnosis tasks take a long time to solve while others require much less time. Intuitively, it would seem to relate to problem complexity. This led to an investigation of alternative measures of complexity of fault diagnosis tasks (Rouse and Rouse, 1979).

A study of the literature of complexity led to the development of four candidate measures which were evaluated using the data from experiments three and five. It was found that two particular measures, one based on information theory and the other based on the number of relevant relationships within the problem, were reasonably good predictors (Pearson's r = 0.84) of human performance in terms of time to solve TASK 1 and TASK 2 problems. The success of these measures appeared to be explained by the idea that they incorporated the human's understanding of the problem and specific solution strategy as well as the properties of the problem itself. Thus, complexity should be viewed as related to both the problem and problem solver.

B. Fuzzy Set Model

One can look at the task of fault diagnosis as involving two phases. First, given the set of symptoms, one has to partition the problem into two sets: a feasible set (those components which could be causing the symptoms) and an infeasible set (those components which could not possibly be causing the symptoms). Second, once this partitioning has been performed, one has to choose a member of the feasible set for testing. When one obtains the test result, then the problem is repartitioned, with the feasible set hopefully becoming smaller. This process of partitioning and testing continues until the fault has been localized and the problem is therefore solved.

If one views such a description of fault diagnosis from a purely technical point of view, then it is quite straightforward. Components either can or cannot be feasible solutions and the test choice can be made using some variation of the half-split technique. However, from a behavioral point of view, the process is not so clear cut.

Humans have considerable difficulty in making simple yes/no decisions about the feasibility of each component. If asked whether or not two components, which are distant from each other, can possibly affect each other, a human might prefer to respond "probably not" or "perhaps" or "maybe."

This inability to make strict partitions when solving complex problems can be represented using the theory of fuzzy sets (Rouse, 1980, 1983a). Quite briefly, this theory allows one to define components as having membership grades between 0.0 and 1.0 in the various sets of interest. Then, one can employ logical operations such as intersection, union, and complement to perform the partitioning process. Membership functions can be used to assign membership grades as a function of some independent variable that relates components (e.g., "psychological distance"). Then, free parameters within the membership functions can be used to match the performance of the model and the human. The resulting parameters can then be used to develop behavioral interpretations of the results of various experimental manipulations.

Such a model was developed and compared to the results of experiments one, two, and four in terms of average number of tests to correctly diagnose faults in TASK 1 and TASK 2 (Rouse, 1978b, 1979b). For TASK 1, the model and subjects differed by an average of only 5%. For TASK

2, with the exception of one trial where two of the subjects made many errors, the comparison was comparable.

Two particularly important conclusions were reached on the basis of this modeling effort. First, the benefit of computer aiding lies in its ability to make full use of 1 outputs shown in Figures 1 and 2, which humans tend to greatly under-utilize. Second, the different strategies of subjects in experiment four can be interpreted almost solely in terms of the ways in which they considered the importance of feedback loops.

It is useful to note here that these quite succinct conclusions, and others not discussed here (Rouse, 1978b, 1979b), were made possible by having the model parameters to interpret. The empirical results did not in themselves allow such tight conclusions.

C. Rule-Based Model

While the fuzzy set model has proven useful, one wonders if an even simpler explanation of human problem solving performance would not be satisfactory. With this goal in mind, a second type of model was developed (Pellegrino, 1979; Rouse, Rouse, and Pellegrino, 1980). It is based on a fairly simple idea. Namely, it starts with the assumption that human problem solving involves the use of a set of situation-action rules (or heuristics) from which the human selects, using some type of priority or control structure (Newell and Simon, 1972; Waterman and Hayes-Roth, 1978; Rouse, 1980).

Based on the results of experiments three, five, and six, an ordered set of twelve rules was found that adequately describes TASK 1 performance, in the sense of making tests similar to those of subjects 89% of the time. Using a somewhat looser set of four rules, the match increases to 94%. For TASK 2, a set of five rules results in an 88% match. It was also found that the rank-ordering of the rules was affected by training, with aided training producing the more powerful rank-orderings.

The new insights provided by this model led to the development of a new notion of computer-aided training. Namely, subjects were given immediate feedback about the quality of the rules which the model inferred they were using. They received this feedback after each test they made. Evaluation of this idea within experiment six resulted in the conclusion that rule-based aiding was counterproductive (36% more tests during training and 159% more upon transfer) because subjects tended to misinterpret the quality ratings their tests received. However, it appeared that ratings that indicated unnecessary or otherwise poor tests might be helpful. This hypothesis was tested and found to be true for FAULT in experiment eight.

D.   Fuzzy Rule-Based Model

All of the modeling results noted above were based on problems in-volving TASK 1 and TASK 2. An attempt was made to apply these models, especially the rule-based model, to describe human performance using FAULT. Success was initially limited by what Rasmussen (1981) would call a shift from topographic to symptomatic search strategies. In other words, once subjects shift from a context-free to context-specific situation, they attempt to use rules that map directly from the symptoms to the solution. In many cases, this mapping process can be adequately described by the earlier rule-based model. However, not infrequently it appears that subjects utilize what might be termed highly context-domi-nated rules, perhaps based on their experiences prior to training.

This dichotomy between symptomatic and topographic problem solving was formalized in a fuzzy rule-based model (Rouse and Hunt, 1981; Hunt, 1981; Hunt and Rouse, 1984a). This model first attempts to find familiar patterns among the symptoms of the failure (i.e., among the state variables of the system). If a match is found, symptomatic rules (S-rules) are used to map directly from symptoms to hypothesized failure. If there are no familiar patterns among the state variables, the model uses topographic rules (T-rules) to search the structure (i.e., functional relationships) of the system. The rules chosen are those with highest membership in the fuzzy set of choosable rules which is defined as the intersection of fuzzy sets of recalled, applicable, useful, and simple rules.

This model was evaluated using the data from experiment eight. It was found that the model could exactly match approximately 50% of subjects' actions and utilize the same rules about 70% of the time. The evaluation of the model also provided a clear demonstration of subjects shifting from S-rules to T-rules when an unfamiliar system was encountered.

E.   An Overall Model

All of the models discussed thus far were devised for the express pur-pose of providing direction to the studies with TASK, FAULT, and real equipment. Of course, considerable effort was also invested in attempting to generalize the model formulations. Thus, the fuzzy rule-based model, for example, certainly appears to be widely applicable. However, none of the models discussed earlier here can really be thought of as describing all of human problem solving.

The fifth and last model to be discussed here represents an attempt to synthesize a model capable of describing human problem solving in gen-eral (Rouse, 1983b). This model is based on a thorough review of the prob-lem solving literature and, to a great extent, the four earlier models. The model operates on three levels: 1) recognition and classification, 2) planning, and 3) execution and monitoring.

Recognition and classification is the process whereby humans determine the problem solving situations with which they are involved. Familiar situations may invoke a standard "frame" (Minsky, 1975) while unfamiliar situations may lead to the use of analogies or even basic principles of investigation. Planning may involve the use of familiar "scripts" (Schank and Abelson, 1979) or, if no script is available, require generation of alternatives, imagining of consequences, valuing of consequences, etc. (Johannsen and Rouse, 1979). Execution and monitoring involves the S-rules and T-rules discussed earlier.

The model operates on the above three levels of problem solving by recursively using a single mechanism that is capable of recognizing both patterns of state information and patterns of structural information. By recursively and constantly accessing this single mechanism the model is capable of both hierarchical (Sacerdoti, 1974) and heterarchical (Hayes-Roth and Hayes-Roth, 1979) problem solving. Simultaneous operation on multiple levels also allows the model to pursue multiple goals such as occur in dynamic systems where the problem solver must coordinate both diagnosing the source of the problem and keeping the system operating.

A particularly interesting aspect of this model's behavior, as well as that of humans, is its potential for making errors. The model has two inherent possibilities for causing errors. The first possibility relates to the model's recursive use of a single basic mechanism. As the model recursively invokes this mechanism, it needs a "stack" or some short-term memory for keeping track of where it is and how it got there. If short-term memory is limited, as it is in humans, the model may recurse its way into getting lost or, pursuing tangents from which it never returns. To constrain this phenomenon, it is probably reasonable to assume that lower priority items in the stack are more likely to be lost first. For example, one is more likely to forget one's umbrella than to forget to go to work.

The second possibility for causing errors is the matching of irrelevant or inappropriate patterns. For example, the model, or a human, may be captured by an inappropriate but similar script or S-rule. As a result, the model may pursue an inappropriate path until it suddenly realizes, perhaps too late to be able to recoup, that it has wandered far afield from where it thought it was headed.

The fact that this model has inherent possibilities for making errors, particularly somewhat subtle errors, provides an interesting avenue for evaluating the model. Most models are evaluated in terms of their abilities to achieve the same levels of desired task performance as humans. A much stronger test would involve determining if the model deviates from desired performance in the same way and for the same reasons as humans. The proposed model can potentially be evaluated in this manner.

However, this model has not yet been evaluated. Thus, at this point, it should mainly be viewed as a synthesis of the wide variety of experimental results and models reviewed here. However, considering the breadth of the investigations upon which it is based, including the extensive review of the literature, this model should also be viewed as much more than conjecture. Clearly, the next step should be evaluation of this model in a variety of problem solving domains.

VI.   DISCUSSION AND CONCLUSIONS

The overall results of this program of research roughly fall into three categories:

1.   Results relating to human problem solving abilities
2.   Concepts for training and aiding problem solvers
3.   Implications for the role of humans in failure situations

In this final section of this chapter, the findings in these three areas will be reviewed.

A.   Human Problem Solving Abilities

Humans are not optimal problem solvers, although they are rational and usually systematic. In general, their deviation from optimality is related to how well they understand the problem, rather than being solely related to properties of the problem. More specifically, suboptimality appears to be due to a lack of awareness (or inability when forced-paced) of the full implications of available information. For example, humans have a great deal of difficulty utilizing information about what has *not* failed in order to reduce the size of the feasible set.

Human problem solving tends to be context-dominated with familiar, or even marginally familiar, patterns of contextual cues prevailing in most problem solving. Humans can, however, successfully deal with unfamiliar problem solving situations, which is a clear indication that human problem solving skills cannot be totally context-specific. Their degree of success with unfamiliar problems depends on their abilities to transition from state-oriented to structure-oriented problem solving. Humans' abilities in the latter mode are highly related to their rank-ordering of rules rather than simply the number of rules available.

Thus, humans' cognitive abilities for problem solving are definitely limited. However, humans are exquisite pattern recognizers and can cope reasonable well with ill-defined and ambiguous problem solving situations. These abilities are very important in many real life fault diagnosis tasks. What are needed, then, are methods for overcoming humans' cognitive limitations in order to be able to take advantage of humans' cognitive abilities.

B. Concepts for Training and Aiding

Throughout this program of research, a variety of schemes have emerged for helping humans to overcome the limitations summarized above. These schemes have been evaluated both as aids during problem solving and as training methods, with evaluation occurring upon transfer to situations without the aids. As noted in previous sections, three types of aid were developed and evaluated.

The first type of aid was implemented within TASK and uses the structure of the network to determine the full implications of the symptoms, as well as each test, with respect to reduction of the size of the feasible set. Basically, this aid is a bookkeeper that does not utilize any information which subjects do not have; it just consistently takes full advantage of this information.

The second type of aid was also implemented within TASK. It evaluates each action by subjects, as they occur, and provides reinforcement in proportion to the degree to which the action is consistent with a context-free optimal strategy. For erroneous (i.e., non-productive) actions, subjects receive feedback that simply notes, but does not explain, their errors. For inefficient (i.e., productive but far from optimal) actions, subjects receive feedback denoting their choices as poor or fair. Optimal or near optimal actions yield feedback indicating the choices to be good or excellent.

The third type of aid was implemented in FAULT. This aid monitors subjects' actions and checks for context-free inferential errors (i.e., errors in the sense of not using the structure of the FAULT network to infer membership in the feasible set). While the aiding is context-free, it explains the nature of the error within the context of the problem (i.e., in terms of the structural implications of the previous actions taken). Thus, the feedback received by subjects not only indicates the occurrence of an error, but also includes a context-specific explanation of why an error has been detected.

The first and third types of aid can both be viewed as structure-oriented bookkeeping aids, while the second type of aid is more strategy-oriented. The results of evaluating these aids were quite clear. The bookkeeping methods consistently improved performance, both while they were available and upon transfer to unaided problem solving. The strategy-oriented aid degraded performance and resulted in negative transfer of training, providing clear evidence of the hazards of only reinforcing optimal performance.

In studies involving transfer from aided TASK to unaided TASK, aided TASK to unaided FAULT, and aided FAULT to unaided FAULT and unaided TASK, positive transfer of training was usually found with the effects most pronounced for unfamiliar systems and fine-grained performance measures. Thus, the evidence is quite clear that humans can be trained to have context-free problem solving skills that, at least partially, help them to overcome the limitations discussed earlier in this section.

Considering transfer from TASK and/or FAULT to real equipment, the results show that training based on simulations such as TASK and FAULT are competitive with traditional instruction, even when traditional instruction provides explicit solution procedures for the failures to be encountered. However, the issue is not really one of TASK versus FAULT versus traditional instruction. The important question is how these training technologies should be combined to provide a "mixed-fidelity" training program that capitalizes on the advantages of each technology (Rouse, 1982b). This mixed-fidelity approach can provide trainees with problem solving principles as well as procedures. Also, it can result in a re-ordering of rules and not just the acquisition of more rules. Thus, this approach can also help humans to overcome the previously discussed limitations. Finally, the mixed-fidelity approach can lead to cost savings since a training program need not rely solely on high-fidelity training devices.

Somewhat as a by-product of this research, a considerable amount was learned about evaluation of training programs (Rouse, 1982a). Perhaps surprisingly, most evaluation efforts in the past have limited consideration to whether or not trainees learn to use the training technology successfully. Few studies have focused on transfer out of the training environment, and even fewer have looked at long-term on-the-job performance. Two of the studies reported here concentrated on transfer to real equipment; a study currently being planned will emphasize on-the-job performance.

One of the key aspects of evaluation is the definition of performance measures. The series of studies reviewed in this chapter began with the use of rather global measures and evolved to the use of very fine-grained measures where, for example, human error was classified using six general and thirty-one specific categories (Rouse and Rouse, 1983). It appears that this detailed level of analysis is very necessary if inadequacies in training programs are to be identified and remedied.

Finally, it should be noted that the model-based approach adopted for these investigations appears to have been a crucial element in their success. The evolving set of models provided succinct interpretations of results and, consequently, generated very crisp hypotheses which focused subsequent investigations. Further, the models contributed to building an overall conceptual view of human problem solving.

C.   The Role of Humans in Failure Situations

Based on the foregoing review of tasks, performance measures, experiments, and models, it seems reasonable to conclude with a discussion of the implications of these results for defining the role of humans in failure situations. As noted in the Introduction, there appears to be a tradeoff between the benefits of humans' unique abilities and the cost of their limitations. Resolving this tradeoff is tantamount to defining the role of humans.

One approach to dealing with this issue is to attempt to automate all fault diagnosis. Unfortunately, what this leads to is automation of routine diagnostic tasks and the human having responsibility for the more difficult problems. As a result, humans perform diagnostic tasks much less frequently; however, when humans must perform the diagnosis, the problem is likely to be very difficult, perhaps even involving untangling of the results of abortive attempts of the computer to diagnose the failures. This is a clear violation of good human factors engineering design principles.

A more appropriate approach is to emphasize computer aiding rather than computerizing. Results reported here indicate that computers can aid humans during training in terms of enhancing general problem solving skills and, during diagnosis by performing bookkeeping functions and monitoring actions to assure that choices are productive. This approach leads to a perspective of humans controlling the problem solving process with sophisticated computer systems providing assistance. As a result, system designers can take advantage of human abilities while avoiding the effects of human limitations.

ACKNOWLEDGMENTS

This research was mainly supported by the U.S. Army Research Institute for the Behavioral and Social Sciences under Contract No. MDA-903-79-C-0421 (1979–1982) and Grant No. DAHC 19-78-G-0011 (1978–1979). Some of the earlier work reported here was supported by the National Aeronautics and Space Administration under Ames Grant No. NSG-2119 (1976–1978). The authors gratefully acknowledge the many contributions of their colleagues William B. Johnson, Sandra H. Rouse, Richard L. Henneman, and Susan J. Pellegrino.

NOTES

1. In the earlier experiments, subjects were not allowed to continue if their choice was incorrect; in the later experiments, they were instructed to continue until the failure was found.

2. For a review of the literature on models of human problem solving, especially for detection, diagnosis, and compensation for system failures, see Rouse (1983b).

REFERENCES

Brooke, J. B., Duncan, K. D. and Cooper, C. Interactive instruction in solving fault finding problems. *International Journal of Man-Machine Studies. 12:*217–227, 1980.

Duncan, K. D. and Gray, M. J. An evaluation of a fault finding training course for refinery operators. *Journal of Occupational Psychology 48:*199–218, 1975.

Hayes-Roth, B. and Hayes-Roth, F. A cognitive model of planning. *Cognitive Science 3:*275–310, 1979.

Henneman, R. L. Measures of human performance in fault diagnosis tasks. Urbana, IL: University of Illinois at Urbana-Champaign, M.S.I.E. Thesis, 1981.

Henneman, R. L. and Rouse, W. B. Measures of human performance in fault diagnosis tasks. *IEEE Transactions on Systems, Man, and Cybernetics SMC-14*, 1984.

Hunt, R. M. A study of transfer of training from context-free to context-specific fault diagnosis tasks. Urbana, IL: University of Illinois at Urbana-Champaign, M.S.I.E. Thesis, 1979.

Hunt, R. M. Human pattern recognition and information seeking in simulated fault diagnosis tasks. Urbana, IL: University of Illinois at Urbana-Champaign, Ph.D. Thesis, 1981.

Hunt, R. M. and Rouse, W. B. Problem solving skills of maintenance trainees in diagnosing faults in simulated powerplants. *Human Factors 23*(3):317–328, 1981.

Hunt, R. M. and Rouse, W. B. A fuzzy rule-based model of human problem solving. *IEEE Transactions on Systems, Man, and Cybernetics SMC-14*, 1984a.

Hunt, R. M. and Rouse, W. B. Computer-aided fault diagnosis training. Submitted for publication, 1984b.

Hunt, R. M., Henneman, R. L. and Rouse, W. B. Characterizing the development of human expertise in simulated fault diagnosis tasks. *Proceedings of the International Conference on Cybernetics and Society*, Atlanta, October, 1981, pp. 369–374.

Johannsen, G. and Rouse, W. B. Mathematical concepts for modeling human behavior in complex man-machine systems. *Human Factors 21*(**6**):733–747, 1979.

Johnson, W. B. Computer simulations in fault diagnosis training: An empirical study of learning transfer from simulation to live system performance. Urbana, IL: University of Illinois at Urbana-Champaign, Ph.D. Thesis, 1980.

Johnson, W. B. and Rouse, W. B. Analysis and classification of human errors in troubleshooting live aircraft powerplants. *IEEE Transactions On Systems, Man, and Cybernetics, SMC-12*, (3):389–393, 1982a.

Johnson, W. B. and Rouse, W. B. Training maintenance technicians for troubleshooting: Two experiments with computer simulations. *Human Factors 24* (3):271–176, 1982b.

Kagan, J. Inidividual differences in the resolution of response uncertainty. *Journal of Personality and Social Psychology 2*(2):154–160, 1965.

Minsky, M. A framework for representing knowledge. In P. H. Winston (Ed.), *The Psychology of Computer Vision*. New York: McGraw-Hill, 1975.

Newell, A. and Simon, H. A. *Human Problem Solving*. Englewood Cliffs, NJ: Prentice-Hall, 1972.

Pellegrino, S. J. Modeling test sequences chosen by humans in fault diagnosis tasks. Urbana, IL: University of Illinois at Urbana-Champaign, M.S.I.E. Thesis, 1979.

Rasmussen, J. Models of mental strategies in process plant diagnosis. In J. Rasmussen and W. B. Rouse (Eds.), *Human Detection and Diagnosis of System Failures*. New York: Plenum Press, 1981, pp. 241–248.

Rasmussen, J. and Rouse, W. B. (Eds.) *Human Detection and Diagnosis of System Failures*. New York: Plenum Press, 1981.

Rouse, S. H. and Rouse, W. B. Cognitive style as a correlate of human problem solving performance in fault diagnosis tasks. *IEEE Transactions on Systems, Man, and Cybernetics SMC-12*(5):649–652, 1982.

Rouse, W. B. Human problem solving performance in a fault diagnosis tasks. *IEEE Transactions on Systems, Man, and Cybernetics SMC-8(4):258–271, 1978a.*

Rouse, W. B. A model of human decision making in a fault diagnosis task. IEEE Transactions on Systems, Man, and Cybernetics SMC-8(5):357–361, 1978b.

Rouse, W. B. Problem solving performance of maintenance trainees in a fault diagnosis task. *Human Factors 21*(2):195–203, 1979a.

Rouse, W. B. A model of human decision making in fault diagnosis tasks that include feedback and redundancy. *IEEE Transactions on Systems, Man, and Cybernetics SMC-9*(4):237–241, 1979b.

Rouse, W. B. Problem solving performance of first semester maintenance trainees in two fault diagnosis tasks. *Human Factors 21*(5):611–618, 1979c.

Rouse, W. B. *Systems Engineering Models of Human-Machine Interaction*. New York: North-Holland, 1980.

Rouse, W. B. Design, implementation, and evaluation of approaches to improving maintenance through training. *Proceedings of the Design for Maintainers Conference*, Pensacola, March, 1982a.

Rouse, W. B. A mixed-fidelity approach to technical training. *Journal of Educational Technology Systems 11*(2):103–115, 1982b.

Rouse, W. B. Fuzzy models of human problem solving. In P. O. Wang (Ed.), *Advances in Fuzzy Set Theory and Applications*. New York: Plenum Press, 1983a, pp. 377–386.

Rouse, W. B. Models of human problem solving: Detection, diagnosis, and compensation for systems failures. *Automatica 19*, 1983b.

Rouse, W. B. and Hunt, R. M. A fuzzy rule-based model of human problem solving in fault diagnosis tasks. *Proceedings of the Eighth Triennial World Congress of IFAC*, Kyoto, Japan, 1981.

Rouse, W. B. and Rouse, S. H. Measures of complexity of fault diagnosis tasks. *IEEE Transactions on Systems, Man, and Cybernetics SMC-9*(11):720–727, 1979. (Also appears in B. Curtis (Ed.), *Human Factors in Softward Development*. Silver Spring, MD: IEEE Computer Society Press, 1981.)

Rouse, W. B. and Rouse, S. H. Analysis and classification of human error. *IEEE Transactions on Systems, Man, and Cybernetics SMC-13*(4):539–599, 1983.

Rouse, W. B., Rouse, S. H. and Pellegrino, S. J. A rule-based model of human problem solving performance in fault diagnosis tasks. *IEEE Transactions on Systems, Man, and Cybernetics SMC-10*(7):366–376, 1980. (Also appears in B. Curtis, (Ed.), *Human Factors in Softward Development*, Silver Spring, MD: IEEE Computer Society Press, 1981.)

Sacerdoti, E. D. Planning in a hierarchy of abstraction spaces. *Artificial Intelligence 5*:115–135, 1974.

Schank, R. C. and Abelson, R. P. *Scripts, Plans, Goals, and Understanding*. Hillsdale, NJ: Lawrence Erlbaum, 1977.

van Eekhout, J. M. and Rouse, W. B. Human errors in detection, diagnosis, and compensation for failures in the engine control room of a supertanker. *IEEE Transactions on Systems, Man and Cybernetics SMC-11*(12):813–816, 1981.

Waterman, D. A. and Hayes-Roth, F. (Eds.) *Pattern-Directed Inference Systems*. New York: Academic Press, 1978.

Witkin, H. A., Ottman, P. K., Raskin, E. and Karp, S. A. *A Manual for the Embedded Figures Test*. Palo Alto, CA: Consulting Psychologists Press, Inc., 1971.

KNOWLEDGE REPRESENTATION AND MAN-MACHINE DIALOGUE

Andrew P. Sage and Adolfo Lagomasino

ABSTRACT

The importance of knowledge in man-machine problem solving tasks, such as fault detection, diagnosis, and correction is well recognized. A meta theory of knowledge is very important in enabling development and use of knowledge based decision support systems by people with diverse experiential familiarity with a particular task. This experiential familarity will strongly influence the method of knowledge representation; cognitive operations on the knowledge base; as well as the memory, control, and user-system interfaces that are most appropriate for a given task. This chapter presents a description and interpretation of several approaches for knowledge representation and knowledge aggregation that support holistic, heuristic, and wholistic reasoning in man-machine systems. Also discussed are issues relative to the design of user interfaces for man-machine dialogue that facilitate use of these three types of reasoning. The need to address uncertainty and imprecision in representing knowledge in man-machine problem solving task is emphasized.

Advances in Man-Machine Systems Research, Volume 1, pages 223–260.

I. INTRODUCTION

There are at least two major complementary viewpoints concerning the design of intelligent man-machine systems. The distinction between the *declarative* and the *procedural* representations (Nilsson, 1980) of knowledge has been an important subject for artificial intelligence and for cognitive science in general. The first of these perspectives emerges from the need to resolve specific issues, or to accomplish specific tasks. This is the declarativist perspective which is typified by direct lines of inferential reasoning using very domain-specific heuristics. It is based on the assumption that "knowledge" consists primarily of "knowing facts," that is to say static, encyclopedic, data-base type knowledge about specific events, objects, and other "elements" and the "relationships" between them. From this viewpoint, the system designer addressed the problem of building machines or systems which exhibit "intelligence" and which are based on explicit knowledge of a specific, generally rather restricted subject area. The expert is, in effect, represented by a stored program, or procedure, with knowledge represented by the explicit content, or data-base, that is imbedded into these programs.

The top level goal of system design, from this viewpoint, is to produce a system capable of assistance that would be regarded as intelligent if the same very specific assistance, based on knowledge of a very explicit set of "standard operating policy" type facts, were obtained from a human. An advantage to declarativist approaches is the directness of the inference chains. This enables them to be especially suited for "concrete operational" mechanical tasks, such that intelligent robots and other forms of automation are generally based on these approaches. A disadvantage to this perspective is that judgments may be made concerning issues where the "expert" is not really expert; and this is not recognized due to the unquestioned use of the facts stored in the data-base, and the generally restricted "formal" reasoning ability that is incorporated into the knowledge based system.

The second viewpoint is aimed at understanding intelligence from the perspective of "knowing how" to use knowledge. This is known as the proceduralist perspective (Winograd, 1975). It is concerned with how to find the most relevant facts from a broader set, and how to make inductive and deductive inferences from them. Thus it is concerned with "formal" knowledge that may be represented by rules concerning "how." Thus, procedural knowledge includes information that enables manipulation and selection from a broad base of declarative knowledge, those aspects deemed most relevant to a given situation, or to achieving a given objective.

Some researchers will also include *control* knowledge as a third component of knowledge that enables coordination of a problem solving task through a variety of processes, structures, and strategies. From this view, performance is based on *skills*, which corresponds to declarative performance, *rules*, which corresponds to procedural performance; and *knowledge*, which corresponds to control performance (Rasmussen, 1983). A somewhat analogous characterization of judgement uses the term wholistic, heuristic, and holistic (Sage, 1981a). Others will simply use the word procedures, or *cognitive engine*, to refer to both procedural and control aspects of problem solving.

Purely declarativist systems are not based on the belief that knowledge is inherently based on procedures for its use. Thus, a purely declarativist approach would seem to be closely related to "wholistic" or "concrete operational" thought in which the response to a given stimulus arises instantaneously out of the experiential data-base of the problem solver without a conscious effort at rule following (Sage, 1981a and references therein). This analogy is not exact, however, as the declarativist perspective is insufficient to describe such wholistic thought processes as intuitive affect and reasoning by analogy. Representation of wholistic thought processes such as these is a present challenge for knowledge based systems (Winston, 1980; Carbonell, 1983).

Proceduralist approaches are rather flexible and effective since a single piece of data will typically be used for multiple purposes. Systems based on this perspective are understandable, and therefore explicable, since they allow for explanation of judgements. They are useful as learning systems since they are capable of explanation. Because of the formal thought process that they involve, adaptation and growth of knowledge over time are possible. Often, however, knowledge based systems designed from this perspective only will not be efficient, or economic in terms of the time required to reach judgments, due to the time required to process a formal approach.

In actual practice, it would appear that the most successful knowledge based systems will be those that use features of both the proceduralist and the declarativist approaches, or which blend concrete operational thought processes with formal operational thought processes as appropriate for the particular issue under consideration and the experiential familiarity of the "expert" with the issue and the environment into which it is imbedded. This is especially the case since it is difficult to conceive of situations in which a totally declarative, or wholistic, or totally procedural, or holistic, perspective is most appropriate as the perspective from which judgment *should* arise. Rather, it would appear that expert "intelligence" and "judgement" will more often come from a general set of procedures for manipulating facts, and a typically large set of facts which describe the expert's knowledge about the issue under consideration, and related issues. A especially important related issue is the experiential familiarity with the environment and the task requirements as imbedded into this environment. This determines the way is which a *cognitive engine* and a *knowledge base* will be invoked so as to enable general procedures to be applied to specific data in order to enable inferences and deductions to be made.

Early efforts in constructing intelligent systems have concentrated primarily on various functional views of intelligence. The results of these efforts have been ad hoc systems that perform successfully in, and perhpas near, the domain for which they were initially constructed. Systems based on the, more global, information processing view that incorporates proceduralist and declarativist perspectives have perhaps been less successful than systems based on the exclusively ad hoc or functional approach. A major difficulty in using a combined approach, however, is that little is known about how the components that comprise a cognitive system interact to enable contingency based processing and use of information in problem solving activities. There are a number of reasons to believe that systems created from this viewpoint are, to a much greater degree than functional systems, domain independent and flexible. As a consequence of this, they should be capable of operation in a much greater variety of environments. Therefore, they should be able to outperform the exclusively functional or ad hoc systems.

Contemporary approaches to knowledge representation typically utilize an information processing approach which allows consideration of the operational components which, at least substantively, comprise an intelligent system (Newell and Simon, 1972, Norman, 1981, Bossel, 1977). These operational components typically include:

1. *Perceptors*—used to receive information,
2. *Effectors*—used to perform actions such as communications,
3. *Representation scheme*—used to interpret and identify information,
4. *Control*—used to monitor and regulate the actions to be performed,
5. *Decisionmaking*—used to allocate cognitive resources
6. *Domain specific Knowledge*—the "facts" that describe a situation,
7. *Meta-knowledge*—Information about knowledge representation (how we know what we know),
8. *Memory*—a physical embodiment of the knowledge base, and
9. *World*—A model of the environment

All of these components are essential for adequate intelligent interaction. No claim is made that this model is physiologically correct, but rather that each of these components is necessary for a substantive representation of intelligence.

There are a number of other models of human information processing and associated judgement. Wohl (1981), for example, has developed a model of human judgement processes that involves *S*timulus information processing, *H*ypothesis generation and evaluation, *O*ption generation and evaluation, and execution *R*esponse. This *SHOR* model is very useful as a framework for structuring decision tasks. Janis and Mann (1977) and Janis (1983) have postulated a stress based model that is very useful in describing how various contingencies lead people to decide how to decide. This model and related models are described in Sage (1981a).

Limits associated with the cognitive capacity of the human mind, time limitations, and many other competing concerns of the decisionmaker are constraints that affect adequate formulation, analysis, and interpretation of complex large scale issues. A design goal for intelligent systems and processes that will assist in various problem solving tasks is to reduce, to the extent possible, the effects of the aforementioned constraints so as to enable the efficient and effective use of information that will lead to quality judgments in routine and familiar task environments and in unfamiliar task conditions.

Several intelligent systems design complexities arise at the cognitive process level of systems management. These relate to the forms, frames, or perspectives associated with acquiring, integrating, and applying vast amounts of knowledge. These forms range from the systemic framework of formulation, analysis and intepretation that is characteristic of formal operational and holistic thought, to intuitive affect that is characteristic of concrete operational and wholistic thought. The knowledge perspectives that are used for a given task depend upon the task requirements, the experiential familiarity of the decisionmaker with the task, and the rationality perspectives that are used for task resolution (Sage, 1981a, 1982; Linstone, 1981).

An appropriate framework in which knowledge could be organized and utilized efficiently and effectively is desired. This is especially needed as studies have shown that the way in which a task is framed exerts a very strong influence upon the way in which task requirements, and task resolution efforts, are determined (Kahneman and Tversky, 1979; Tversky and Kahneman, 1981). This requires that we be able to address the modeling of knowledge based decision support systems from several perspectives; especially as these relate to the different components of an intelligent system. Of particular interest will be those components at the interface between the cognitive process level of systems management and the problem level, and at the knowledge metalevel which will enable ef-

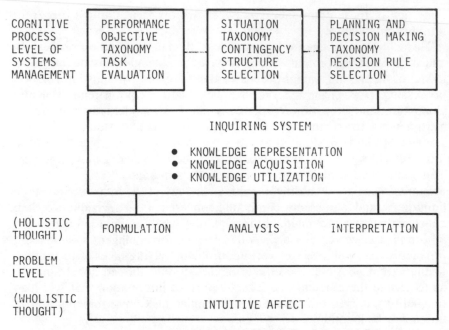

Figure 1.   Interface Between Metalevel Concerns of Systems
Management and Problem Solving Activities

fective modeling of the cognitive system itself. There seems little question
but that this will involve an inquiring system (Churchman, 1971), as suggested in the conceptual diagram of Figure 1.

In this chapter, we will expand upon these notions by next discussing
several approaches for the representation of knowledge in intelligent systems. Then we will examine some meta-knowledge considerations that
are very important in the design of intelligent systems. Following this,
we indicate some concerns relative to the veridicality of expert bases.
We argue that there exists a major need for incorporation of approaches
that enable questions of uncertainty and imprecision to be considered.
Several approaches that may be useful towards this end are described.
The chapter concludes with a discussion of research needs in this area.

II.   KNOWLEDGE REPRESENTATION IN INTELLIGENT
SYSTEMS

Approaches that will enable effective knowledge representation, and associated inference activities, in large knowledge bases have been the subject of investigation for many researchers. Perhaps the most complete
listing of current work on various approaches to this subject appears in

the Special Issue of *SIGART* on Knowledge Representation edited by Brachman and Smith (1980). The information provided for each project include a list of personnel, the address, a brief description of the project, and references to literature where more detailed description can be found. Although there are several other representations the definitive *Handbook of Artificial Intelligence* (Barr, Cohen, and Feigenbaum, Vol 1, 1981) describes seven representation schemes: logic, procedural representations, semantic networks, production systems, direct (analogical) representations, semantic primitives, and frames and scripts. Each of these may be used to describe the four different types of factual knowledge elements that may be captured in a knowledge base: objects, events, performance, and meta-knowledge. These representations will also assist in identification of the values that need to be associated with facts in order to enable judgement and choice. Finally, they may be used to describe knowledge retrieval, reasoning, and acquisition of new knowledge and relating this to already known knowledge. An ultimate goal of all of this is knowledge-based program construction (Barstow, 1979).

Newell (1982), in discussing the role of knowledge and its representation, has introduced the concept of a knowledge level mechanism that is analogous both to the concept of computer systems levels, as well as to the information processing view of intelligent systems that we have briefly described. Knowledge at the knowledge level is perceived as the medium by which an agent fulfills its goals and/or explains its actions. Goals and actions are the essential components at the knowledge level. Finally, the behavioral law that controls this mechanism is the principle of rationality which states that actions are selected to attain the agent's goals. This principle implies goal optimization or satisfaction in the attainment of target aspirations.

The main theme of Newell's research is that knowledge and its representation, although highly related, exist at two different levels. Knowledge exists as the *medium* at the knowledge level described above, and its representation lies at the *program* or *symbol* level in a computer system. This implies the existence of a knowledge base, a cognitive engine, and appropriate control and interface mechanisms to enable communication between these elements of the knowledge base system and the system input and output. This representation is not unlike the representation model for the problem space and associated problem solving activities in traditional means-ends analysis (Newell and Simon, 1972). In the means-ends representation of the problem space, there exists:

1. a *problem space*;
2. a set of all possible problem states within the problem space;
3. one state which is known as the *initial problem state*;

4. one, or possible more than one, state which is known as the *goal state*;
5. A set of *conditional operators* that will transform one problem state into another problem state, all within the problem space;
6. an *error comparator* which computes the difference between two states, typically between the goal state and the present state;
7. a *controller* which applies a control that is a function of the error difference that is detected; and
8. a set of path constraints that must be satisfied in order for a problem solution to be admissible.

Within this problem space, problem solving using means-ends analysis is comprised of four generic activities:

1. comparison of the current state with the goal state;
2. choice of a control that will reduce this difference;
3. application of this control if it is admissible and, if it is not, determination of a suitable subproblem and applications of means-ends analysis to the subproblem; and
4. resumption of effort on the original problem or task when the subproblem has been solved.

It is from this perspective that the issue of knowledge representation will be addressed in this chapter. Approaching knowledge representation from this perspective will also help in the understanding of some of the many meta-knowledge concerns that are required to ensure acceptable human-machine interactions. Although we will not explore the point in any detail here, there is much present evidence to show that analogous reasoning (Silverman, 1983) is *not* necessarily a totally wholistic process; and that at least some forms of analogous reasoning can be represented by holistic processes, such as means-ends analysis (Carbonnel, 1983).

As we have noted, our concern here will be primarily with the information processing representation of knowledge. The purpose of a particular knowledge representation is to enable the use of knowledge for: retrieving factual information from the knowledge base that is judged relevant to the task at hand, reasoning about these facts in the search for a resolution of the task requirements, and acquiring more knowledge.

A.   Production Systems

There are various mechanisms that can be used to represent the organization of declarative and procedural knowledge. The most common, and simplest, representation is that in which knowledge is structured as a set of facts, with each fact related to one or more facts in a causal type of inferential relationship. This modular cause-effect structuring of knowledge into the form of a production rule, which was first developed by Newell and Simon (1972), has seen many applications including strategic planning, policy analysis, decision making and other areas. Many of these are in areas formally outside the artificial intelligence communities (Axelrod, 1976; Roberts, 1976; Eden et al., 1979; Mitroff et al., 1982) and employ constructs that are more restrictive than those that are often found in production rule systems. We will comment on some of these approaches to knowledge representation and machine learning (Davis and King, 1977; Davis and Buchanan, 1977; Davis, Buchanan, and Shortliffe, 1977; Michalski, Carbonell, and Mitchell, 1983) later in this chapter.

The basic idea behind a production system is that there exists a set of productions, or rules, in the form of various condition-action pairs, generally in the form of "IF THEN" combinations. Initially these were exclusively explicit rules, although there is much current interest in incorporating fuzzyness and imprecision into production rule concepts (Hunt, 1982; Rouse, 1983). Generally production rules are heuristic in nature, and may be appropriate or inappropriate to the task at hand. The normative goal of an expert system is to use "good" heuristics, of course.

B.   Semantic Networks

The problem of representing knowledge in terms of *nodes*, that represent objects concepts or events, and *links* between the nodes to represent their interrelations, is one of the forms of knowledge representation that has been of continuous interest in the field of artificial intelligence. Much of the initial research concerning *semantic networks*, as these representations are called, stems from the desire to model human associative memory. Current research (Winograd, 1983) is very concerned with the representation of strings in natural language for story understanding (Norman, Rumelhart, et al., 1975; Schank and Abelson, 1977; Findler, 1979; Carbonnel, 1981).

The work on semantic network representation, such as represented in the works of Quillian (1968), Schank (1975, 1982), Rumelhart and Ortony (1977) and others, has been especially concerned with basic notions for representing human verbal knowledge in "understander" systems. Schank and his colleagues, for example, have been especially concerned with the use of semantic networks as aids in the teaching of reading. Semantic networks are usually described in terms of their purpose, such as the purpose of aiding reading or aiding the understanding of the belief structure of a person concerning some issue.

The basic concept of having nodes that represent elements in the universe and links that represent the contextual relations between these elements is very appealing. A semantic network is intended to represent concepts expressed by natural-language words and phrases as interconnected nodes, connected by a particular set of arcs called semantic relations. Concepts in this system of semantic networks are word-sense meanings. Semantic relations are those which the verb of a sentence has with its subject, object, and prepositional phrase arguments; in addition to those that underlie common lexical, classificational, and modificational relations. It would appear that the coding method of Wrightson in Axelrod's work on cognitive mapping (1976), which is based on Axelrod's earlier definitive structural modeling work and the theory of psycho-logic (1972), and Eden's cognitive mapping coding techniques (1979), are based on these sentence characteristics.

A semantic network is a convenient computational representation, that is readily implementable on computers, in which to represent knowledge which is expressed in natural language. The extent to which this type of knowledge representation can be effectively utilized for inference processes depends on the existence, or lack thereof, of a well structured and sophisticated set of rules. The antecedent-consequent type of rule is commonly used. Structurally the left-hand side of the rule, or antecedent, represents the set of assumptions or conditions appropriate for use of the rule. The right-hand or consequent side of the rule represents the set of end results. Conjunctions, disjunctions, and negations can occur on either side of the typical relational rule. This form of representation, although simple, has been powerful enough to support much of the early work in artificial intelligence. The contemporary frame and script concept, in which the structure and framework within which new information is interpreted in terms of the concepts that have been acquired through previous experience, has evolved from semantic network concepts. Again we see the strong contextual dependency, or expectation driven processing, of scripts and frames such that one looks for things based on the context that one believes exists.

Semantic networks are capable of representing both the physical descriptions of actions as well as the purposeful aspects of these actions. Thus, a semantic network representational system must include the *goals* or objectives that can be obtained by actions, the *scripts* which describe scenarios in simple stereotyped situations, the *plans* which allow for flexible description of potentially appropriate action-impact pairs, and the *themes* which allow for such environmental descriptions as the occupations and aspirations of actors involved in the issue under consideration. In this way, a semantic network may be descriptive as well as a prescriptive mechanism.

Unfortunately, semantic networks do have a number of defects. Winston and Brown (1979) have cited three of these:

1. they lack a way of smoothly creating aggregate concepts that can be manipulated as simply as they could be if they were elementary concepts;
2. they lack a simple mechanism whereby one concept can acquire or otherwise inherit information from another concept, other than by those that have been explicitly stated in the inference rules; and
3. they lack a structural representation that is internal to a given concept.

There are also bound to be difficulties when the size of the semantic network, in terms of the associated data-base, becomes sufficiently large such that it can represent a nontrivial amount of knowledge. The computational difficulties in processing the network and the cognitive difficulties in coping with the associated complexity may be overwhelming. This leads to the need for network aggregation in order to obtain "summary" representations that are efficient and effective. These aggregate networks are called *frames*. Thus the concept of a frame, originally due to Minsky (1975), as a chunk of slots and their contents does eliminate many of the defects of semantic networks.

There are also other difficulties associated with the semantic network concept. There are, for example, difficulties associated with representing time, as structural models are basically static devices. There are difficulties associated with maintaining a distinction between "facts" and "values," and in incorporating concepts of uncertainty, fuzzyness, and imprecision. Situations often arise, for example, in which reasoning must be accomplished using incomplete, imprecise, and perhaps even contradictory data. These may arise when collecting data from information summaries, or from imperfect and/or distributed sources. Much research is being done in semantic networks to improve their usefulness as representational schemes for general knowledge. A collection of papers edited by Findler (1979) describes various applications, recent developments, and extensions of semantic networks. Some applications illustrate the use of semantic networks in constructing the knowledge base of programs which exhibit some aspects of understanding—paraphrasing, abstracting, and classifying a corpus of text, answering questions on the basis of common sense reasoning, drawing deductive and inductive inferences, and obeying commands.

C. Cognitive Maps

A much simpler network based representation is one in which knowledge is structured as a set of concepts; with each concept related to one or more facts by a single causal type of relationship. Needed to accomplish a structural model of this sort is a theory of *psycho-logic* (Abelson and Rosenberg, 1958; Abelson, 1973; Axelrod, 1972) or pulsed digraphs (Roberts, 1976) in which relations may have enhancing ($+1$), inhibiting (-1), or neutral (0) causal influences on other relations. A number of applications of the resulting *cognitive maps*, as the resulting structural models are often called, have been reported by Axelrod (1976), Eden et al. (1979) and others. Very early uses of the term cognitive map described how animals cognized the spatial environment around them (Tolman, 1948), and this particular restrictive use of the term is still common (Downs and Stea, 1977). In its more general form, a cognitive map is the result of an attempt to capture an individual's view of the world with respect to a particular issue.

Unlike semantic networks, that are typically multi-relational structures and as such require a sophisticated set of production rules or grammars for representation and interpretation, a cognitive map is based on a single specific contextual relation, such that any given element will have enhancing, inhibiting or neutral causal impacts on each other element in the cognitive map. Thus the representation and analysis of a cognitive map will usually be simpler than is the case for semantic networks. The simple contextuality of the cognitive map may make it difficult for such a map to replicate a complex belief structure. However, the elements which represent concepts in a cognitive map are variables that can take on different values. The linkage, or contextual relation, among concepts may represent causal assertions and perceptions concerning how one concept variable affects another concept variable.

Since the dynamics of the reasoning mechanism in a cognitive map is imbedded in structural considerations, it is a simple matter to simulate the reasoning process of a person if we can assume that the cognitive map has been faithfully constructed. Nozicka, Bonham, and Shapiro (1976) and Eden et al. (1980) have developed computer simulations of cognitive map constructs using graph theoretic methods. These methods provide a convenient matrix technique for the representation and manipulation of concepts and their structural relationships (Harary, et al., 1965; Sage, 1977). Among the operations that may be better understood and communicated by means of a cognitive map are the formulation of alternative explanations of an event that has occured or is anticipated to occur (diagnosis), the development of perceptions concerning the expected consequences of an event (prognosis), and the search and ranking of relevant policy options (decisionmaking). It is necessary, of course to be aware of the considerable possibilities for flawed judgment in these activities (Sage and White, 1983), and a major use for a cognitive map may well be to explore the possibilities for flawed information processing and judgement that may result from use of such a map.

D. Frame Representation

Minsky (1975), in a paper on frame-systems, presents a rather different approach to semantic networks that eliminates some of their defects. He advocates the use of *local procedures* within a "frame" in order to represent structured knowledge. A frame is, as we have noted, a chunk of knowledge for representing a stereotyped situation. Attached to each frame are several kinds of information. Some of this information concerns how to use the frame and some of the information may concern what can happen next. A frame can then be represented as a hierarchical network of nodes and relations. The top level element in the frame will represent facts that are always assumed to be true about the generic situation at hand. The lower levels of the frame will have many empty terminal slots that must be filled by the specific context dependent information about the frame and the situation at hand.

Although this frame-based theory attempts to address identification of a general system, whose purpose is to represent knowledge as a collection of separate and simple fragments, there seems to exist many technical gaps at present concerning how to design an operational system that makes best use of this theory of frames. Automated procedures to organize knowledge in such a hierarchically structured framework appear necessary to make this theory functional. The problem of efficient search involving a very large knowledge base is one of great concern both in terms of time, and in terms of the possible combinatorial explosion of knowledge that can occur. Some authors (Hayes-Roth and Waterman, 1978; Bobrow and Winograd, 1977; Bobrow and Collins, 1975; Wilensky, 1981; Davis and Lenat, 1982) have subscribed to the idea of using meta-knowledge to guide the search procedures in order to reduce the resulting knowledge based system to a manageable size. Others, such as Fahlman (1979), have advocated parallel processing through networks of structural knowledge in order to reduce the search time in very large knowledge-based systems as much as possible.

E. Representating Analogies

As we have noted previously, analogies and analogous inference play a very important role in human judgment. Philosophers of science often claim that reasoning by analogy is the basic for hypothesis formation and identification in science. Often analogous reasoning (Sternberg, 1977, 1982; Gick and Holyoak, 1980; Winston, 1980) is used when there exists uncertainty and imprecision associated with the judgment task at hand. We have alluded to learning as a process whereby we are able to do things more efficiently, or effectively, or more efficiently and effectively the next time that we do them. Thus learning can involve rote memory and direct implementation of instruction, to reasoning by analogy, and through discovery and observation, of a concrete or formal nature.

When there is a lack of explicit, certain, and concrete knowledge about a specific issue, reasoning by analogy will be often used. In such cases, one searches for similarities between the task situation extant and a previously experienced and familiar situation. When the situations are sufficiently analogous that one can see "parallels" between elements in one situation and elements in the other, then reasoning by analogy becomes possible.

Silverman (1983), in an extensive review of research concerning analogous inference in systems management, identifies a taxonomy that facilitates the development of procedures and protocols for the identification and correction of pitfalls in this form of reasoning. Sternberg (1977) also presents a descriptive theory of analogous reasoning and experimentally identifies sources of errors that may occur during each of the operations that constitutes his paradigm for analogous reasoning. The basic operations in Sternberg's model are: encoding, inference, mapping, application, and response. This research provides a great deal of descriptive information that has motivated the development of normative theories of concept formation and issue formulation. Carbonell (1983) has modeled analogous reasoning using means-ends analysis. He uses this model to integrate skill refinement with knowledge acquisition as an effective procedure for reliable and effective problem solution.

Nakamura and Iwai (1982) and Nakamura, Sage and Iwai (1983) have developed a questioning-answering system for information retrieval in which the system relates its own associative knowledge with the user's knowledge in an analogous fashion. The system's associative knowledge is represented as a semantic network and an information metric is introduced to measure topological distances that represent analogous reasoning similarities. The question-answering dialog is such as to encourage user responses which enable identification of analogous situations through direction of the questioning to enable maximum similarity determination.

F.   Summary

This section has been concerned with discussing those aspects of knowledge representation that seem most relevant to the overall objectives of this chapter. By way of summary, a knowledge representation is a set of symbols used for illustration, a method for arranging them, and a reasoning mechanism for using the arrange symbols to hold and convey knowledge. Sound approaches to knowledge representation are needed as the foundation for much contemporary large systems efforts in the human-machine systems area, regardless of whether the application efforts involve, fault diagnosis and repair, planning, language understanding, or any of a large number of areas.

There is no available theory comparing the different types of knowledge representation schemes that is capable of indicating which will be the most useful in any particular application. Nevertheless, several authors including McCarthy (1969), and Winston (1977) have suggested different and useful criteria that are very important in enabling selection from among the several knowledge representation schemes. These criteria include:

1. *epistemological adequacy*—sufficient knowledge should be present and capable of being captured by the knowledge representation scheme used such that the task requirements can, in principle, be met,
2. *heuristic adequacy*—a knowledge representation approach potentially expressing information helpful in solving problems also should offer ways of avoiding or greatly reducing search complexity, and
3. *extendability*—the knowledge based system should be designed such as to minimize the difficulty of associating and linking new information to existing information.

Finally, we might remark that it is especially necessary that the knowledge representation scheme be capable of coping with the types of expertise, and the reasoning perspectives, that can reasonably be expected to exist among the users of the knowledge based system. In this way, we will be much better able to accomplish needed activities that involve learning and discovery (Hayes-Roth, 1983) such as the diagnosis of faulty theories, the proper assimilation of new knowledge, and useful frameworks in which to pose questions such that they are understandable and interpretable in the way in which the questioner (should have) intended. Clearly, all of this has major implications for subjects such as decision-making in general (Ungson and Braunstein, 1982) and such important subareas as human detection and diagnosis of system failures (Rasmussen and Rouse, 1981).

III. SOME META-KNOWLEDGE CONSIDERATIONS IN THE DESIGN OF INTELLIGENT SYSTEMS

Various assumptions about the nature and characteristic of the contingency task structural elements of task, environment and human problem solver familiarity with these are considered essential in the design of inquiry systems in order to enable effective and efficient organization of knowledge about specific situation domains. Among these assumptions, the following are especially pertinent here:

1. The world is basically orderly enough so that, at least imperfect, observation of it is possible.
2. Identification of the task and the environment may, generally, not be performed in a value free context.
3. Some value judgments always preceed the collection of any set of data, or the construction of any model.

There are two basic components that comprise any nontrivial inquiry system such that a combination of information structures and appropriate interpretation procedures will lead to intelligent behavior: a knowledge base that will contain relevant facts concerning objects, events, performance, and contain a value system as well; and a cognitive engine that will enable aggregation of the facts and values such as to enable judgements. Each of these is needed in an inquiry system in order that it be possible to use knowledge through the acts of *information retrieval, reasoning* using facts and values, and *acquisition and aggregation* of new knowledge which relates something that is new relative to that which is already known. There exist two diametrically opposed perspectives with respect to learning, or acquisition of new knowledge and subsequent aggregation of new knowledge into an existing knowledge base. At this point we could make a distinction between the knowledge acts of learning and discovery, and the manner in which these perspectives differ according to whether knowledge acquisition is through learning or discovery. This would involve us in concerns of passive knowledge acquisition versus active, or experiential, knowledge acquisition. The journey would be interesting, unfortunately somewhat long, and not among the most important of concerns that are of interest to us here. Sternberg's (1982) definitive work contains much detail associated with concerns of this nature.

A. Knowledge Acquisition: Learning

The two perspectives referred to in the foregoing concern how individuals go about, in a procedural manner, the retrieval of information, the use of information for reasoning, and the feedback process that enables acquisition of new knowledge. One perspective concerning this is that learning is performed by an elementary-to-complex process in which simple things are learned first and from this, more advanced concepts are then learned. The other perspective is based on the belief that learning starts with complex statements about the description of a situation; and that through decomposition into simpler statements a system is able to increase its understanding concerning the specific situation domain.

As a means of illustrating the two perspectives concerning the knowledge acquisition process, it is useful to compare problem solving activities and natural language understanding in some detail with respect to these two perspectives concerning learning or acquisition of new knowledge.

The elementary-to-complex process of learning could be characterized by activities involved in understanding natural language. For example, natural language statements are generally composed of *actors, actions*, and a set of *cases* that are associated with particular actions. From this, the reader must infer why the action has occurred and what things must have been present in order for the action to be possible to occur. To aid *understanding* of a story, the reader searches for an explanation of the action or situation described in the story.

Problem solving, on the other hand, involves the construction of a plan to satisfy a set of goals. It generally involves a process of iterative use, and reuse, of understanding mechanisms. Often, especially when the problem solver does not recognize an inherent structure to the problem such that wholistic thought is possible, this involves a process of decomposition or disaggregation of task components in a manner that is typical of the complex-to-elementary mode of knowledge acquisition. The process of decomposing a problem into manageable subgoals is clearly an important aspect of planning and problem solving, at least for initially unstructured or semistructured problems.

The analysis of Sacerdoti (1977) is concerned with this process. His top down planning system, NOAH, is based on a knowledge structure that permits subgoals to be chosen and placed in an efficient sequence. This is accomplished through use of a large number of small modular programs, each of which may contain specialized knowledge about actions and their impacts on the problem domain, which may be invoked at times that are appropriate to the goals that the actions will achieve.

At one extreme in the knowledge aggregation role, learning is represented by inputs of a set of facts and inference rules about a specific domain. The system has no control over this domain and cannot question the validity of judgements that result from the use of facts and inference rules in it. At the other extreme, judgements evolve from "facts," "values," and an aggregation procedure for facts and inference rules that can be questioned and tested with respect to simplicity and truth through use of the existing accumulated knowledge in the knowledge base. These processes may involve the identification of inconsistencies in the aggregated knowledge base and resolution, and associated efforts to correct these by means of some well structured man-machine dialogue.

Most artificial intelligence systems that are in use today are based on the first perspective with respect to acquisition and aggregation of new knowledge. That is to say, they use the elementary-to-complex perspective with respect to knowledge acquisition; and also do not verify and validate, or otherwise seek to determine the consistency of, the resulting knowledge base. The resulting lack of control with respect to questions of validity of the resulting knowledge base is characteristic of an incomplete intelligent system. If this perspective of knowledge acquisition and aggregation is used exclusively, then some essential components of an intelligent cognitive system have been omitted, or the interaction between the knowledge base and the user has been modeled inadequately.

As previously noted, systems that operate at the other extreme of the knowledge acquisition spectrum, such that learning proceeds in a complex-to-elementary fashion, have been difficult to implement. In reality, both modes of learning are appropriate and both are used by the human problem solver. Although the knowledge base component of decision support systems is generally small, the typical knowledge base of a management oriented decision support system is often based on complex-to-elementary elicitation of subjective knowledge. Integration of the two approaches is clearly desirable. We will briefly address this topic and the related topic of veridicality in knowledge based systems in Section IV.

B.   Information Seeking: Input

The representation of knowledge at the symbol level suggests the existence of some form of prior knowledge which enables the system to perform the function of acquisition and aggregation of the new knowledge in with the existing knowledge. We have discussed, in the previous section, various approaches to knowledge representation in terms of their characteristics, functions, and purposes. We will now describe a general knowledge representation system in terms of some elementary apriori knowledge that is assumed common to any representational scheme. This apriori knowledge is characterized by the existence of an input component capable of labeling, chunking, structuring, and storing data into the resulting knowledge base.

The input component is concerned with the processes by which information that is relevant to a situation is obtained from the environment. The three basic cognitive processes involved in this are perception, consciousness, and memory. Together, these provide both a description and an explanation of the situation such as to enable generation of a set of *beliefs* or *knowledge*, organized into a "representation," about the situation. It should be noted that what may be a belief to one person may be knowledge to another. Abelson (1979) presents a cogent discussion of this subjective interpretation of belief and knowledge. There must also be a generalization component, or inference mechanism that is equipped with some form of basic logic to enable access to the knowledge base for formation of inferences and judgments. A fundamental question arises from this discussion concerning how apriori knowledge and the generalization component influence the operation of the complete knowledge representation system.

In the context of devices to aid in human information processing, there exist two design philosophies concerning the proper relationship between the input component and the generalizing sector. One advocates the view that these sectors should be considered as operating separately; with the input component in charge of knowledge acquisition and representation considered as if it was independent of the generalizing sector which is in charge of aggregation. This approach offers simplicity in system design, at the expense of effectiveness, as the mode of representation of information will influence the success or failure of the inquirer in arriving at a solution. This deficiency seems to be characteristic of most current information systems. They are passive systems and it is up to the user to recognize an information need and then seek out the required information.

The other approach considers that the two sectors are essentially non-separable and that each supports and enhances the functioning of the other; but this in turn complicates the system design, perhaps by a considerable amount. In this approach, the internal interactions of the input and generalizing sector are capable of generating user-system interaction. There are various ways in which a system can initiate a dialog with a user (Michie, 1980):

1. identification of "gaps" in the knowledge base that prevent the system from making inferences, or from adequately summarizing the information in a sector of the knowledge base;
2. identification of an inconsistent set of information followed by detection of the inability of the system to resolve it; and
3. inability of the system to satisfy the desired goals of the user.

Identification of these potential deficiencies and use of prompts based on them for purposes of computer-control dialog are needs in intelligent system design. The systems *AM* and *TEIRESIAS* described in Davis and Lenat (1982) are state of the art programs that perform system initiated dialog of the first type described above. AM, for example, is designed to reason about a set of existing elementary knowledge and to create new concepts and plausible hypotheses based on this knowledge base. The knowledge base that is used to test the veridicality of the system is a set of mathematical propositions and heuristics that are commonly used in mathematics, the subject domain for AM. In this process, the system searches, by means of internal and external communications, for new concepts and/or plausible heuristics with which to identify, generate, or "discover" new concepts. Lenat notes that AM has not discovered anything new to the body of mathematical knowledge, but has been able to provide interesting interpretations to well known concepts in mathematics. Perhaps most importantly, it has demonstrated that a symbiotic man-machine combination is potentially able to produce better results than either might do alone (Fischer, 1983).

TEIRESIAS is designed to allow interactive capture and transfer of expertise from a human expert to the knowledge base of a system, using dialog that is generated and initiated by the system. The system is capable of improving its knowledge base by identifying defects in the existing base, a *debugging-acquisition* phase, as well as explaining the why and the how of the conclusions that it reaches, that is to say an *explanation phase*. The similarity between these two systems is their capacity for reasoning and conducting inquiry on the basis of knowing about the system's own knowledge.

In at least one way, incomplete and inconsistent knowledge is valuable as the recognition of this may encourage people, and knowledge based systems, to initiate questioning as part of a search for potentially disconfirming information. Mitroff et al. (1982) describe a system that conducts inquiry on the basis of defects, such as these, in the knowledge base. In their system, defects may arise in either of two ways: through improper aggregation of information from distributed, potentially conflicting, information sources; or through the internal generation of conflicting premises that lead to challenges to the system user concerning the veridicality and completeness of the existing knowledge base.

This last mode of system initiated and generated dialog involves a process that is the inverse of the sequence normally followed in means-ends analysis (Simon and Newell, 1972). This is essentially an *inverse optimal control problem* (Sage and White, 1977), or a problem for which the regression analysis based *policy capture* (Hammond et al., 1980) is appropriate. It is also related to the *Hegelian inquirer* perspective described by Churchman (1971) that serves as the basis for the dialectic inquiry approach that is advocated by Mason and Mitroff (1981), Mitroff and Mason (1982), and Mitroff, Mason, and Barabba (1982). Approaches to directing inquiry into the structure of decision situations (Sage and Rajala, 1978; Rajala and Sage, 1980a,b; Pearl et al., 1982) in terms of the response to queries to the decisionmaker are related to this approach also.

In the definitive goal directed structuring system (*GODDESS*) of Pearl, for example, the inquiry process proceeds from the identification of desired goals to actions that might achieve the goals, and then conditions that must exist in order for these actions to be optimal in achieving the goals. In turn, these conditions generate more subgoals. It has been shown

(Pearl et al. 1982) that the goal directed dialog stimulates the generation of relevant ideas and provides for the structuring of initially ill structured problems in a more effective and efficient manner than undirected questioning. External user-system interaction of the sort that we have described here is a highly desired feature for information systems. It should, in principle, strongly influence the integrated functioning and the success of the input and generalizing sectors. Doubtlessly future knowledge based system designs will be much influenced by these considerations.

IV. CONCERNS RELATIVE TO VERDICALITY OF EXPERT KNOWLEDGE BASES

In the past, it has often been accepted as an unchallenged truth that the judgment of an expert is better than that of the layperson. Presumably this occurs because the cognitive processing of the expert is better, perhaps due to much prior experience and learning, than the novices. Much recent work has shown that the judgment of experts is often quite flawed. This is especially the case when the contingency task structure has changed from the one in which the expert is truly an expert, and this change is not recognized. Recent reviews of this work, including references to many of the studies that have been performed may be found in Kahneman, Slovic, and Tversky (1982) and Sage (1981a).

Judgements must, at least in a substantive fashion, result from the integration of facts and values relative to some task requirement or issue. Many studies show that judgements can be explained by the way a person has developed a scheme, schema, representation, or mental framework that enables interpretation of the situation and resulting task requirement or issue. A person sees the world with the aid of an individual mental framework; the world around people is generally too complicated for them to perceive every part of it and every possible combination of parts. Therefore, a simplified framework is necessary to relieve the problem solver of a very difficult cognitive burden, and allow representation, and structuring of issues and determination of the requirements for judgment and choice relative to the task at hand. Many studies have shown that this framework exerts a very strong influence over the identified task requirements as well as the impacts that the problem solver is likely to perceive as resulting from alternative options.

Errors in the process of judgment and choice may result when the mental framework is flawed or inadequate for the situation at hand. These errors might result from: the use of poor heuristics for integration of facts and values; biases associated with the incomplete acquisition, that is selective perception, of information; biases associated with the failure to seek potentially disconfirming information, flawed structuring of the information that is acquired; and use of poor heuristics to analyze information flow within an otherwise correct framework. Also the value system may be inchoate or labile. It is not feasible to present a detailed listing of the many possible information processing biases here. Some of the major conclusions of studies relevant to this topic are:

1.   People often make very incorrect judgments about random events when they trust intuition rather than formal inference analysis. Small sample individuating experiences are often, for example, believed to be far more highly representative of a population than they should be based on statistical theory. Intuitions about random samples indicates that there is a strong belief in a law of small numbers which indicates that very small samples are highly representative of a population, often more highly representative than large samples.

2.   Errors in judgment due to a variety of flaws are often more associated with cognitive information processing limitations than they are to direct intentions to mislead. Intuitive feelings are often due to unintentional but consistent misperceptions and nonopportunistic wishful thinking.

3.   Normatively based intelligent systems which require intuitive (expert) judgment about facts may well be subject to a variety of cognitive information processing biases and associated judgmental errors unless they are cognizant of the strong implications for expert system design of this recent work.

All of this suggests much caution in the use of unaided intuitive judgment in the construction of knowledge bases for expert systems. It might be argued that the typical expert, especially the scientific expert, uses quantitative or qualitative analysis procedures in the process of judgment and choice. But, there are a number of studies (Kahneman, Slovic, and Tversky, 1982; Dreyfus and Dreyfus, 1980; Dreyfus, 1982; Klein, 1980, 1982) which suggest that this view is not correct; that experts act very intuitively in the formulation of concepts and hypotheses, and in the concept and hypotheses interpretation efforts that lead to judgment and choice. Often these intuitive wholistic processes are sound. However when the environment, or more generally the contingency task structure, has changed and this change is not recognized, then the lack of suitable prior experience with the new situation may well mean that the use of formal operational and holistic thought is more appropriate than intuitive wholism as there is no relevant knowledge base that truly justifies the decisionmaker's belief of mastery of the situation at hand.

It has been shown that many people, including even experts, search very selectively for that information which confirms their existing beliefs and hypotheses. They selectively ignore information which is potentially disconfirming to their existing beliefs and hypotheses. Often beliefs are based upon a primacy effect in which information occurring early in time is used to form inferences and allows data saturation such that new information is, in effect, not processed. This allows once acquired beliefs to persevere long after the evidence, on which the beliefs were based, has been thoroughly discredited. Several explanations are possible. Beliefs become part of the value system through a sublimation process and this results in an emotional commitment to "facts" that have become part of the value system. Inability to accomplish double loop learning (Argyris, 1974, 1978, 1982) results in actual theories of action that are inconsistent with espoused theories of action. These are not observed by the person responsible for judgement, although it is often noticed by others who do not inform the responsible person, and this further complicates the situation.

It would appear that efforts to examine the potential for elicitation and encoding of flawed heuristics and cognitive biases in knowledge based systems is a need at this time. This is needed such that corrective measures may be undertaken. Of considerably greater normative significance would be the development of design procedures and protocols for expert based systems that will minimize, to the extent possible, the presence of flawed heuristics and cognitive biases in both the accumulated knowledge base as well as in the inference mechanism or cognitive engine that constitutes a knowledge based system. To do this effectively will require very careful attention to the many uses for information (Feldman and March, 1981; Keen, 1981) in addition to the traditional systems engineering and management science use as data of value for decisionmaking.

It may be argued that our discussions in the preceeding sections indicate that the knowledge base and the associated inference mechanism in an expert consulting system will be highly structured and analytic, and that sound rational inferential relations will be used to process information that establishes the knowledge base. This is generally correct if only because of the fact that knowledge representation in a computer must, at least at this time, be of a formal nature. The knowledge that is "captured" from experts may well, and generally will, be of an intuitive and wholistic form initially. To represent the knowledge of many experts in an appropriate form in a computer appears to require that the represented knowledge be, translated perhaps, in a formal structured form. Later, we will argue that this will require approaches to uncertainty, fuzzyness, and imprecision that are only now being explored.

Correct quantitative representation of the parameters within an assumed but potentially flawed framework is only assurance that the analysis portion of the task of judgment and choice is veridical, conditioned upon proper framing or formulation of the decision situation. In terms of intelligent systems, this insures that this conditional concept analysis is sound. But there is no assurance at all that concept framing, or formulation of the issue, is sound. As a consequence of this, the impacts from alternative courses of action may well not be properly determined due to poor formulation and structuring of the decision situation. The effects of this upon the interpretation process, which involves the aggregation of facts and values by a user of the intelligent system, will be difficult to predict although they will generally result in suboptimum behavior, perhaps behavior that is considerably suboptimal to that which would have resulted from a totally intuitive process.

Multiple representations of conceptual information, representations which may or may not be in conflict with one another, are another way in which to approach the issue of verdicality in knowledge based systems. This is, of course, based upon the belief that there will somehow be ultimate truth contained in the union of all of these multiple perspectives. This leads to difficulties, however, in that the selection of that subset of the information contained in a multiple perspectives approach which represents "truth" may be extraordinarily difficult. This is especially the case when multiple responses, each presented from a different perspective, from a single query contain conflicting and non-commensurate information. We will address some potential efforts to resolve these difficulties later in our next section.

V.   METHODS FOR REPRESENTING IMPRECISE KNOWLEDGE

In this section we will describe approaches to knowledge representation that are based in the form of the messages used to convey knowledge and to relate its representation with the imprecision in judgment that is common in natural language discourse. First, we will characterize approaches based on the theory of fuzzy sets. Then we will look at some methods based on plausible reasoning.

A.   Fuzzy Set Theory

It is necessary to introduce some notation related to the axiomatic basis for fuzzy sets. We say that a message is represented by means of a statement composed of a sequence of elements that convey knowledge when the elements in the statements are given an interpretation on a specific domain or universe of discourse. Thus, a statement may be represented abstractly as $A_1^n(x_1, x_2, \ldots, x_n)$, where $x_1, x_2, \ldots x_n$ represent the elements of the statement, A_1^n is the predicate, and where n is the degree of the predicate.

There are three types of statements of primary importance for knowledge representation by means of fuzzy sets. The classification of a statement is based on the form of the statement. We have the following.

1. *Predication*—for n = 1, we have a predicational or classificational type of statement. $A_1^1(x)$ denotes then that x, the element, is classified as a member of the set described by the monadic predicate A_1^1.

2. *Relational*—for n > 1, we have a relational statement. $A_1^n(x_1, x_2, \ldots x_n)$ denotes that a relation defined by A_1^n, among the elements $x_1, \ldots x_n$, exists.

3. *Functional*—a special type of relational statement of the form $A_1^n(x_1, \ldots x_n)$, where n > 1. When A_1^n and all but one of the x's are specified, the value of the remaining x is completely determined. If $A_1^2(x_1, x_2)$ is a function, then $x_2 = A_1^2(x_1)$ is computable by knowledge of A_1^2 and x_1. Statements of this form acquire meaning when the elements and predicates that form the statements are associated with well defined elements in a specific domain or universe of discourse. This is called an interpretation and it is said that an expression is an interpretable statement. A potential difficulty that is present in representation of schemes described so far, concerns the way in which the validity of the expressions stored in the knowledge base, in the form of facts and inference rules, is represented.

In classical mathematical logic, the use of a two-valued "true-false" level of discourse is generally adequate for analyzing the foundations of mathematical thought. The traditional two-valued system or even multivalued systems have been felt by many to be inadequate to model analytically the verbal modes of human communication that are due primarily to imprecision and uncertainty.

In response to this perceived inadequacy of classical mathematical logic, Zadeh (1965) proposed the idea of fuzzy sets. This and several other seminal papers described how imprecision and uncertainty in judgement, generally present in the use of natural languages, can be modeled through a simple generalization of the classical definition of a set of elements. A classical set S may be represented by a membership function f_s, by letting $f_s(x) = 1$ if x is in the set S and letting $f_s(x) = 0$ if x is not in S. The notion of a fuzzy set is implemented by allowing f_s to range over the entire interval from 0 to 1. Thus, one obtains the idea of the degree or grade of membership in S. As an example, if T is the fuzzy set of tall persons, then a person who is six feet in height might have .9 as a grade of membership in T, while a person who is five feet six inches in height might have a grade of membership of only .4 in T. The assignment of such grades of membership is, by and large, the product of a subjective interpretation in the context of a given domain or universe of discourse. That is, a building which is only six feet in height in the fuzzy set of tall buildings would have a very low grade of membership.

To appreciate the generalization of fuzzy sets in terms of its capability in modelling imprecision, it is benefical to compare the process of interpreting a sentence in both classical logic and fuzzy logic. An interpretation is the act of giving meaning to a statement in any language. To interpret a statement, a domain or universe of discourse must be defined. A domain D is a set that may consist, for example, of the set H of human beings and a well defined relation R_1 like "is the father of" among members of the set H. In this example, the relation R_1 is a subset of the 2-fold Cartesian product of H, $H \times H$. Suppose that x and y correspond to John and Peter, respectively, and each are members of the set of human beings, and A_1^2 a predicate function which corresponds to the relation "is the father of" in the domain of human beings H_1. Then the expression $A_1^2(x,y)$ takes on the interpretation "John is the father of Peter."

Associated with each expression on a specific domain is a valuation system that assigns values as a measure of the verdicality of the given expression. In the case of the foregoing example, the truth value of the expression can be described as

$$A_1^2 = \begin{cases} \text{TRUE} & \text{if } A_1^2(x, y) \in R_1 \\ \text{FALSE} & \text{if } A_1^2(x, y) \in R_1 \end{cases}$$

In general, a valuation is a mapping of the set of expressions into the set of values of the valuation system. A two-valued logic was sufficient to describe the validity of the expression in the above example. However, there are expressions that may make it more convenient to use the generalization of classical sets of elements, that we have described previously, into fuzzy sets. The set of values of the valuation system is extended to a generalized multivalued logic in which the verdicality of an expression in a given domain is perhaps represented by, say, a fuzzy subset of [0,1]. This might be the fuzzy representation used instead of the points in [0,1]; as is characteristic of most multivalued logics, where 1 indicates absolute truth of the expression and 0 a definite denial of the expression (Zadeh, 1975a,b).

As an illustration, consider the symbolic expression $A_1^1(x)$, where, for purpose of interpretation, x corresponds to a member of the set of human beings, say {JOHN}, and the monadic predicate A_1^1 represents "is a young person." Then, the statement $A_1^1(x)$ meaning "John is a young person" may be associated with a truth value that is a subset of the range [0,1]. The particular subset chosen would be selected according to the age of the person and the, perhaps subjective, interpretation of the meaning of this age variable in the context of the given domain of young persons.

Assuming that the diagramatic representation in Figure 2 is adequate to describe the degree of membership, then the valuation of the expression $A_1{}^1(x)$ could be assessed by knowing the age of x.

Fuzzy sets have been used in the interpretation of other types of imprecision commonly found in natural language dialogues. In particular, the approach has been used to characterize "linguistic hedges" or modal qualifiers that strengthen or weaken the meaning of sentences. Typical modal qualifiers are "very," "more or less," "quiet," "almost" etc. The result of using a linguistic hedge in a sentence is a shift in its relative "truth" value for a given interpretation. "John is tall" may have, for example, a truth value of .3, so the statement "Mike is very tall" should have a greater truth value than the previous statement, and "Charles is not so tall" should have a smaller truth value with respect to the first statement. It is especially important to note the relative contextual dependency of modal qualifiers. There may be a much greater shift ratio in the meaning of "very" in the expression "very fat" than there is in the expression "very tall." For while the weight ratio between normal weight and very fat might be two to one, it would be unreasonable that the height ratio between normal height and very tall would be two to one. Again, this says that we need to have some indication of the anchoring and adjustment implied by modal qualifiers, and this requires a knowledge of context, before we can associate quantitative meaning to these fuzzy expressions.

Zadeh (1973) has introduced a procedure which enables manipulation of linguistic hedges. The procedure is based on an intuitively plausible subjective interpretation of the effect on truth values of the hedges that are used in simple statements. No research appears to have been done with respect to the validity of this interpretation, and its efficacy as a standard procedure for manipulating hedges. However, it is apparent that the subjective nature of this interpretation makes this theory open for

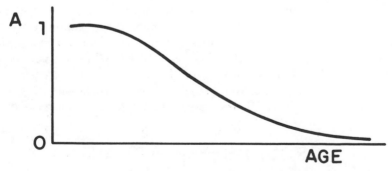

Figure 2. Interpretation of a Fuzzy Membership Function.

criticism, as has been noted in several commentaries, such as that by Schefe (1980).

There have been a number of, primarily conceptual, applications of fuzzy sets as a model of approximate reasoning. These have included such areas as multiple criteria decision making, data base management, and natural language processing (Ruspini, 1982; Gupta and Sanchez, 1982a,b). Even though there have been some very promising initial studies of this important topic (Rouse, 1980) considerable research is still needed in order to assess impacts that fuzzy set theories of approximate reasoning may have in the development and operational implementation of man-machine systems.

B.   Models of Logical Reasoning

In describing the structure of the beliefs and the statements that people make about: issues that are of importance to them, the nature of the environment that surrounds them, as well as the ways in which people reason and draw conclusions about the environment and issues that are imbedded into the environment, especially when there are conflicting pieces of information and opinions concerning these, people often attempt to use one or more of the forms of logical reasoning. There are many important works that deal with this subject including Harre (1970), Wason and Johnson-Laird (1972), Falmagne (1975), Levi (1980), and Tweney et al. (1981). Of particular interest in this connection is the works of Toulmin et al. (1979) in that they have constructed an explicit model of logical reasoning that is subject to analytical inquiry and computer implementation. Further, the model is sufficiently general that it can be used to represent logical reasoning in a number of application areas.

Starting from the assumption that whenever we make a claim there must be some ground in which to base our conclusion, Toulmin states that our thoughts are generally directed, in an inductive manner, from the GROUNDS to the CLAIM as shown in Figure 3. The GROUNDS and the CLAIM, in the context of what we have said thus far, are statements that express facts and values. As a means of explaining observed patterns of stating a claim, there must be a reason, that can be identified, with which to connect the GROUNDS and the CLAIM. This connection is called the WARRANT, and it is the warrant what gives to the GROUNDS-CLAIM connection its logical validity as shown in Figure 4.

We say that the GROUNDS supports the CLAIM on the basis of the

GROUNDS ⟶ CLAIM

Figure 3.   Illustration of the Grounds Supporting the Claim.

Figure 4. Illustration of the Way in Which the Warrant Adds Validity
to the Support of the Grounds to the Claim.

existence of a WARRANT that explains the connection between the
grounds and the claim. It is easy to relate the structure of these basic
elements with the process of inference, whether inductive or deductive,
in classical logic. The WARRANTS are the set of rules of inference, and
the GROUNDS and CLAIM are the set of well defined propositions or
statements. It will be only the sequence and procedures, as used to formulate the three basic elements and their structure in a logical fashion,
that will determine the type of inference that is used.

Sometimes, in the course of reasoning about an issue, it is not enough
that the WARRANT will be the absolute reason to believe the CLAIM
on the basis of the GROUNDS. For that, Toulmin allows for further
BACKING which, in his representation, supports the WARRANT, as
shown in Figure 5.

It is the BACKING that provides for the reliability, in terms of truth,
associated with the use of the WARRANT. The relationship here is analogous to the way in which the GROUNDS support the CLAIM. An argument will be valid and will give the CLAIM solid support only if the
WARRANT is relied upon and is relevant to the particular case under
examination. The concept of logical validity of an argument seems to
imply that we can only make a CLAIM when both the WARRANT and
the GROUNDS are certain. However, imprecision and uncertainty in the
form of exceptions to the rules or low degree of certainty in both the
GROUNDS and the WARRANT does not prevent us on occasions from
making a "hedge" or a vague CLAIM. Very commonly, we must arrive
at conclusions on the basis of something less than perfect evidence; and

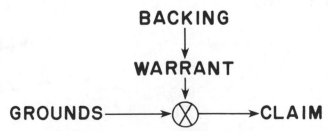

Figure 5. Partial Support Structure for Logical Reasoning.

we put those claims forward not with absolute and irrefutable truth but rather with some doubt or degree of speculation.

To allow for these cases Toulmin adds MODAL QUALIFIERS and POSSIBLE REBUTTALS to his framework for logical reasoning. MODAL QUALIFIERS refer to the strength or weakness with which a claim is made. In essence every argument has a certain modality. Its place in the structure presented so far must reflect the generality of the WARRANTS in connecting the GROUNDS to the CLAIM. POSSIBLE REBUTTALS, on the other hand, are exceptions to the rules. Although MODAL QUALIFIERS serve the purpose of weakening or strengthening the validity of a CLAIM, there may be still conditions that invalidate either the GROUNDS or the WARRANTS, and this will result in deactivating the link between the CLAIM and the GROUNDS. These cases are represented by the POSSIBLE REBUTTALS.

The resulting structure of logical reasoning provides a very useful framework for the study of human information processing activities (Sage, 1981b). It is beneficial to stress at this point that the order in which the six elements of logical reasoning has been presented serve only the purpose of illustrating their function and interdependence in the structure of an argument about a specific issue. It does not represent any normative pattern of argument formation. In fact, due to the dynamic nature of human reasoning, the concept formation and framing that results in a particular structure may occur in different ways. The six element model of logical reasoning is shown in Figure 6.

The effect of various forms of inquiry upon such important issues in human information processing as information requirements determination

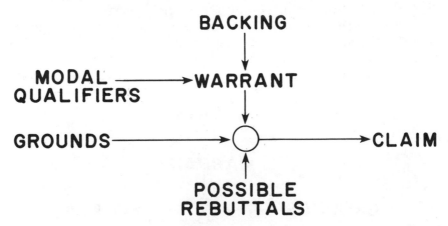

Figure 6.   The Six Element Model of Logical Reasoning.

(Davis, 1982), detection of inconsistencies and information processing biases could potentially be investigated by using this structure of rational argument.

Mason and Mitroff [1981], Mitroff and Mason [1982] and Mitroff, Mason, and Barabba (1982) have utilized Toulmin's structure of arguments together with the theory of plausibility and inconsistency of Rescher and Manor (1970) to develop a framework for handling ill-structured decision problems. An information system to aid in planning and problem solving activities must have an appropriate framework for capturing the structure of a complex issue; and the ability to represent, analyze, and synthesize knowledge from many disparate sources. Generally, when the knowledge from these disparate sources are conjoined, a set of inconsistent or contradictory conclusions follow. This is the viewpoint taken by Mitroff, Mason, and Barabba in their research on policy as argument, based on this model of logical reasoning.

The concept of a maximally consistent subset and plausibility indexing of Rescher and Manor makes it possible to identify and resolve such inconsistencies. A maximally consistent subset is the largest set of propositions that can be formally conjoined without logical contradiction. A proposition could be any of the six elements of logical reasoning in Toulmin's structure. In set theoretical terms, let S be any set of propositions $S = \{P_1, P_2 \ldots\}$. We say that the subsets $S_1, S_2 \ldots$ of S are the maximal consistent subsets (MCS) of S if each satisfies the following:

1. S_i is a nonempty subset of S;
2. S_i is consistent; and
3. No element of S that is not a member of S_i can be added to it without generating an inconsistency.

The basic idea in the foregoing is to identify the MCS and select, on the basis of plausibility indexes computed on the MCS, that subset which represents best the situation extant. The concept of plausibility in this work is similar to the possibility concept of Zadeh. Each author claims that, although close similarities exist with concepts of subjective probability, the concepts differ in that they are based on different axiomatic structures. These views are not universally accepted. The essence of the disagreement seems to be that when we talk of the probabilities of probabilities, or associate uncertainty with uncertainty, and extend this meta-level concept to probabilities of the probabilities of probabilities, then we will ultimately generate a new probability density function. This is essentially the argument of deFinetti (1972), and has been extensively discussed in the works of Sahlin (1983), Goldsmith and Sahlin (1983), Gardenfors and Sahlin (1982).

VI. SUMMARY

A purpose of this chapter has been to discuss knowledge based concerns relative to man-machine problem solving tasks; such as fault detection, diagnosis and correction. These are very important concerns at this time as advances in technology render physiological skills that involve strength and motor abilities relatively less important than they have been, and significantly increase the importance of cognitive and intellectual skills. The need for humans to monitor and maintain the conditions necessary for satisfactory operation of systems, and to cope with the poorly structured and imprecise knowledge that must be brought to bear on unforeseen occurrences, is greater than ever.

A number of advances in electronic technology provide computer and communication systems that enable a significant increase in the amount of data that is available for judgement and decisionmaking tasks. However, even the highest quality data will generally be associated with considerable uncertainty and imprecision.

A large number of studies show that unaided humans are very inadequate at a number of human information processing tasks, and that the associated acts of judgement and choice are often quite flawed. Simply supplying more of the vast amount of information, that is now potentially available, is not the answer as studies show that the presence of greater amounts of unstructured information may simply make the decisionmaker's performance more erratic, due to the use of various forms of selective perception to cope with the vastly increased amount of information and associated cognitive overload. In the face of all of this, the decisionmaker is likely to resort to time honored and tested methods of judgment that are based on wholistic and intuitive behavior. When true expertise is involved, these wholistic and intuitive approaches will often be both the appropriate response, and the effective and efficient one as well. However when the environment, or contingency task structure as a more general descriptor, has changed and this change is not recognized then it is very possible that the former "master" becomes no master at all, except perhaps of the art of self deception.

All of this has major implications with respect to the design of systems for the human user and requires, for appropriate system design, an understanding of human performance in problem solving and decisionmaking tasks. This understanding has to be at a *descriptive* level, such that we can predict what humans will likely do in particular situations, and at a *prescriptive* level, such that we can *aid* humans in various cognitive tasks.

There are many situations in which this need for knowledge based support for the human operator is strong. With respect to man-machine systems, a prototypical task might concern automated systems that require: only occasional minor changes in parameter settings for maintenance of satisfactory performance; and rare major changes in the control inputs of a structural nature in order to prevent significant malfunction, and perhaps disaster. The need for significant interaction with an operational system will often be very infrequent. The advance notice of the need to interact will often be very short, the importance of the consequence of the human-machine interaction will be very large, the need to interact will occur at times that are very unpredictable, and there will be large amounts of information available when the potential need for operator interaction occurs. A specific description of the situation in which a knowledge based system is needed will form one of the bases (Davis, 1982) for the determination of information and task requirements for knowledge base system design. The importance of appropriate knowledge representation to assist the human operator, and the need for suitable man-machine interaction and dialog is apparent. A number of other generic man-machine situations could be cited, that range from strategic planning to manual control of machine tools, and the need for decision support and associated knowledge representation would be easily recognized in each of them. Unfortunately, the activities that need to take place in order to establish the most appropriate knowledge representation and decision support system are not so transparent.

In this chapter, we have discussed forms of knowledge representation that are of value to these ends. The form, or frame, of knowledge representation that a person uses is very much a function of the perspective that the person has with respect to the particular issue under consideration. This suggests a contingency task structural approach as being very important. For it is the particular task at hand, the environment into which this task is imbedded, and the experiential familiarity of the decisionmaker with the task and environment that determines the information acquisition and analysis strategy that is adopted as a precursor to judgment and choice. Thus we need to be aware of a variety of knowledge representations; and the way in which meta-level knowledge leads to a knowledge representation in terms of the information requirements determined for a particular task, the method of analysing the acquired information, and the way in which associated facts and values are aggregated to enable judgement formation.

We have been especially concerned here with the fact that the information that is used for judgement and choice is typically not precise clerical and accounting data, but a mixture of this data and information of an imprecise and uncertain nature. As a consequence of this, we have described approaches based of fuzzy sets and logical reasoning that allow incorporation of notions of imprecision. Much of our own recent work has been in this area as well (White, Sage, and Scherer, 1982; Sage and White, 1984). Needless to say, we believe that it is an important area for continued research, together with the many other activities that are associated with understanding and improving decisions.

ACKNOWLEDGMENT

This research was supported by the United States Army Research Institute for the Behavioral and Social Sciences under Contract MDA 903-82-C-0124.

REFERENCES

Abelson, R. P., and Rosenberg, M. J. Symbolic psycho-logic: A model of attitudinal cognition. *Behavioral Science 3:*1–13, 1958.

Abelson, R. The structure of belief systems. In Schank and Colby, 1973, pp. 287–339.

Abelson, R. Differences between belief and knowledge systems. *Cognitive Science 3:*355–366, 1979.

Argyris, C. Some limits to rational-man organizational theory. In *Systems and Management Science.* New York: McGraw-Hill, 1974, pp. 33–47.

Argyris, C. *Reasoning, Learning and Action.* San Francisco: Jossey Bass, 1982.

Argyris, C., and Schon, D. A., *Theory in Practice: Increasing Professional Effectiveness.* San Francisco: Jossey Bass, 1974.

Argyris, C., and Schon, D. A., *Organizational Learning: A Theory of Action Perspective.* Reading, MA: Addison Wesley, 1978.

Axelrod, R. M. *Framework for a General Theory of Cognition and Choice,* Berkeley: Institute of International Studies, University of California, 1972.

Axelrod, R. M. (Ed.), *Structure of Decision: The Cognitive Maps of Political Elites.* Princeton, NJ: Princeton University Press, 1976.

Barr, A., Cohen, P. R., and Feigenbaum, E. A. (Eds.), *Handbook of Artificial Intelligence,* Vols. I, II, and III. Los Altos, CA: William Kaufman, 1981 and 1982.

Barstow, D. R., *Knowledge-Based Program Construction.* New York: American Elsevier, 1979.

Bobrow, D. G., and Collins, A. (Eds.), *Representation and Understanding: Studies in Cognitive Science.* New York: Academic Press, 1975.

Bobrow, D. G. and Winograd, T. An overview of KRL, a knowledge representation language. *Cognitive Science 1*(1):3–46, 1977.

Bossel, H. (Ed.), *Concepts and Tools of Computer Assisted Policy Analysis,* 3 vols. Stuttgart: Birkhauser, 1977.

Brachman, R. J. and Smith, B. C. Special issue on knowledge representation. *SIGART Newsletter* No. 70, February 1980, Published by Association for Computing Machinery.

Carbonell, J. G. *Subjective Understanding: Computer Models of Belief Systems.* Ann Arbor, MI: UMI Research Press, 1981.

Carbonnel, J. G. Learning by anology: Formulating and generalizing plans from past experience. Chapter 5 in Michalski, R. S., et al., *Machine Learning: An Artificial Intelligence Approach.* Palo Alto, CA: Tioga, 1983, pp. 137–162.

Churchman, C. W. *Design of Inquiring Systems.* New York: Basic Books, 1971.

Davis, G. B. Strategies for information requirements determination. *IBM Systems Journal 21*(1):4–30, 1982.

Davis, R., and Buchanan, B. G. Meta-level knowledge: Overview and applications. *IJCAI 5:*920–927, 1977.

Davis, R., Buchanan, B. G., and Shortliffe, E. H. Production rules as a representation for a knowledge-based consultation system. *Artificial Intelligence 8:*15–45, 1977.

Davis, R., and King, J. J. An overview of production systems. In E. Elcock and D. Michie (Eds.), *Machine Intelligence,* vol. 8. New York: John Wiley, 1977, pp. 300–332.

Davis, R., and Lenat, D. B. *Knowledge-Base Systems in Artificial Intelligence.* New York: McGraw Hill, 1982.

De Finetti, B. *Probability, Induction, and Statistics.* New York: John Wiley, 1972.

Downs, R. M., and Stea, D. *Maps in Minds: Reflections in Cognitive Maps.* New York: Harper & Row, 1977.

Dreyfus, S. E. Formal models vs human situational understanding: Inherent limitations in the modeling of business expertise. *Office: Technology and People 1:*133–165, 1982.

Dreyfus, S. E., and Dreyfus, H. L. A five stage model of the mental activities involved in directed skill acquisition. University of California at Berkeley, Report ORC 80–2, 1980.

Eden, C., Jones, S., and Sims, D. *Thinking in Organizations.* London: The McMillan Press Ltd., 1979.

Eden, C., Smithin, T., and Wiltshire, J. Cognition, simulation and learning. *Journal of Experiential Learning and Simulation 2:*131–143, 1980.

Fahlman, S. E. *NETL: A System for Representing and Using Real-World Knowledge.* Cambridge, MA: MIT Press, 1979.

Falmagne, R. J. (Ed.), *Reasoning: Representation and Process.* Hillsdale, NJ: Erlbaum, 1975.

Feldman, M. S., and March, J. G. Information in organizations as signal and symbol. *Administrative Science Quarterly 26:*171–186, 1981.

Findler, N. V. (Ed.), *Associative Networks: The Representation and Use of Knowledge by Computers.* New York: Academic Press, 1979.

Fischer, G. Symbiotic, knowledge-based computer support systems. *Automatica,* Vol x, No. y, December, 1983.

Gardenfors, P., and Sahlin, N. E. Unreliable probabilities, risk taking, and decision making. *Synthese 42:*00–00, 1983.

Gick, M. L., and Holyoak, K. J. Analogical problem solving. *Cognitive Psychology 12:*306–355, 1980.

Goldsmith, R. W., and Sahlin, N. E. The role of second order probabilities in decision making. A Chapter in Humphreys, et al. (Eds.), *Analysing and Aiding Decision Processes.* Amsterdam: North Holland, 1983.

Gupta, M. M., and Sanchez, M. (Eds.), *Fuzzy Information and Decision Processes.* New York: North Holland, 1982.

Gupta, M. M., and Sanchez, M. (Eds.), *Approximate Reasoning in Decision Analysis.* New York: North Holland, 1982.

Hammond, K. R., McClelland, G. H., and Mumpower, J. *Human Judgment and Decision Making.* New York: Praeger, 1980.

Harary, F., Norman, R., and Cartwright, D. *Structural Models: An Introduction to the Theory of Directed Graphs.* New York: John Wiley, 1965.

Harre, R. *The Principles of Scientific Thinking.* Chicago, 1970.

Hayes-Roth, F. Using proofs and refutations to learn from experience. In Michalski et al., *Machine Learning.* Palo Alto, CA: Tioga, 1983, pp. 221–240.

Hayes-Roth, F., and Waterman, D. Principles of pattern directed inference systems. In Waterman, D., and Hayes-Roth, F. (Eds.), *Pattern Directed Inference Systems.* New York: Academic Press, 1978.

Hunt, R. M. Rule-based models of human control of dynamic processes. *Proceedings 1982 Systems Man and Cybernetics Annual Conference,* October, 1982, pp. 686–689.

Janis, I. L. and Mann, L. *Decision Making.* New York: Free Press, 1977.

Janis, I. L. *Groupthink.* New York: Free Press, 1983.

Johnson-Laird, P. N., and Wason, P. C. *Thinking: Readings in Cognitive Science*. Cambridge: Cambridge University Press, 1977.

Kahneman, D., Slovic, P., and Tversky, A. (Eds.), *Judgment under Uncertainty: Heuristics and Biases*. Cambridge: Cambridge University Press, Cambridge, 1982.

Kahneman, D., and Tversky, A. Prospect theory: An analysis of decisions under risk. *Econometrica 27*(2):263–291, 1979.

Keen, P. G. W. Information systems and organizational change. *Communications of the Association for Computing Machinery 24*(1):24–33, 1981.

Klein, G. A. The use of comparison cases. *Proceedings 1982 Systems Man and Cybernetics Annual Conference*, October 1982, pp. 88–91.

Klein, G. A. Automated aids for the proficient decision maker. *Proc. IEEE Systems Man and Cybernetics Conference*, October 1980, pp. 301–304.

Levi, I. *The Enterprise of Knowledge*. Cambridge, MA: M.I.T. Press, 1980.

Linstone, H. A. et al. The multiple perspective concept. *Technological Forecasting and Social Change 20*:275–325, 1981.

Mason, R. O., and Mitroff, I. I. *Challenging Strategic Planning Assumptions*. New York: John Wiley, 1981.

McCarthy, J. and Hayes, P. J. Some philosophical problems from the standpoint of artificial intelligence. In Meltzer, B. and Michie, D. (Eds.), *Machine Intelligence*, Vol. 4, Edinburgh, Edinburgh University Press, 1969.

Michalski, R. S., Carbonell, J. G., and Mitchell, T. M. *Machine Learning: An Artificial Intelligence Approach*. Palo Alto, CA: Tioga, 1983.

Minsky, M. A framework for representing knowledge. In P. A. Winston (Ed.) *The Psychology of Computer Vision*, New York: McGraw Hill, 1975.

Michie, D. *Knowledge Based Systems*. Department of Computer Science, University of Illinois, Urbana, Report UIUCDCS-R-80-1011, 1980.

Mitroff, I. I., Mason, R. O., and Barabba, V. P. Policy as argument—A logic for ill-structured decision problems. *Management Sciences 28*(12):1391–1404, 1982.

Mitroff, I. I., and Mason, R. O. Business policy and metaphysics: Some philosophical considerations. *Academy of Management Review 7*(3):361–371, 1982.

Nakamura, K., and Iwai, S. Topological fuzzy sets as a quantitative description of analogical inference and its application to questioning-answering systems for information retrieval. *IEEE Transactions on Systems Man and Cybernetics SMC 12*(2):193–204, 1982.

Nakamura, K., Sage, A. P., and Iwai, S. An intelligent database interface using psychological similarity between data. *IEEE Transactions on Systems Man and Cybernetics SMC 13*(4):00–00, 1983.

Newell, A., and Simon, H. A. *Human Problem Solving*. Englewood Cliffs, NJ: Prentice-Hall, 1972.

Newell, A. The knowledge level. *Artificial Intelligence 18*:87–127, 1982.

Nilsson, N. J. *Principles of Artificial Intelligence*. Palo Alto, CA: Tioga, 1980.

Norman, D. A. *Perspectives on Cognitive Sciences*. Hillsdale, NJ: Lawrence Erlbaum, 1981.

Norman, D. A., Rumelhart, D. E., and the LNR Research Group, *Explorations in Cognition*. San Francisco: Freeman, 1975.

Nozicka, G. J., Bonham, G. M., and Shapiro, M. J. Simulation techniques. In Axelrod, R. M., (Ed.), *Structure of Decisions*. Princeton, NJ: Princeton Univ. Press, 1976.

Pearl, J., Leal, A., and Saleh, J. GODDESS: A goal directed decision structuring system. *IEEE Transactions on Pattern Analysis and Machine Recognition PAMI-4*(3):250–262, 1982.

Quillian, M. R. Semantic Memory. In Minsky, M. (Ed.), *Semantic Information Processing*. Cambridge, MA: MIT Press, 1968.

Rajala, D. W., and Sage, A. P. On measures for decision model structuring. *International Journal of Systems Science 11*(1):17–31, 1980a.

Rajala, D. W., and Sage, A. P. On decision situation structural models. *Policy Analysis and Information Science* 4(1):53–81, 1980b.

Rasmussen, J. Skills, rules, and knowledge: Signals, signs, and symbols; and other distinctions in human performance models. *IEEE Transactions on Systems Man and Cybernetics SMC* 13(3):a–b, 1983.

Rasmussen, J., and Rouse, W. B. (Eds.), *Human Detection and Diagnosis of System Failures*. New York: Plenum Press, 1981.

Rescher, N. and Manor, R. On inference from inconsistent premisses. *Theory and Decision* 1:179–217, 1970.

Roberts, F. S. *Discrete Mathematical Models: With Application to Social, Biological, and Environmental Problems*. Englewood Cliffs, NJ: Prentice Hall, 1976.

Rouse, W. B. Models of human problem solving; Detection, diagnosis, and compensation for system failures. *Automatica* Vol. 19, No. 6, December 1983.

Rouse, W. B. *Systems Engineering Models of Human-Machine Interaction*. New York: Elsevier, 1980.

Rumelhart, D. E., and Ortony, A. The representation of knowledge in memory. In Anderson, R. C., Spiro, R., and Montague, W. (Eds.), *Schooling and the Acquisition of Knowledge*. Hillsdale, NJ: Lawrence Erlbaum Associates, Inc., 1977.

Ruspini, E. H. Possibility theory approaches for advanced information systems. *Computer* (September):83–91, 1982.

Sacerdoti, E. D. *A Structure for Plans and Behavior*. New York: Elsevier, 1977.

Sage, A. P., and White, C. C.III. ARIADNE: A knowledge based interactive system for planning and decision support using imprecise parameters and dominance structure elication. *IEEE Transactions on Systems Man and Cybernetics* 14(1), 1984.

Sage, A. P., and White, E. B. Decision and information structures in regret models of judgment and choice. *IEEE Transactions on Systems Man and Cybernetics* 13(2), 1983.

Sage, A. P. Behavioral and organizational models for human decisionmaking. *Proc. IEEE Conference on Decision and Control*, December, 1982.

Sage, A. P. Organizational and behavioral considerations in the design of information systems and processes for planning and decision support. *IEEE Transactions on Systems, Man, and Cybernetics SMC-11*(9):640–678, 1981a.

Sage, A. P. Hierarchical inference in large scale systems. *Proc. IFAC World Congress* 40(August):1–9, 1981b.

Sage, A. P., and Rajala, D. W. On the role of structure in policy analysis and decision making. In J. W. Sutherland (Ed.), *Management Handbook for Public Administration*. New York: Von Nostrand, Reinhold, 1978, pp. 568–606.

Save, A. P. *Methodology for Large Scale Systems*. New York: McGraw Hill, 1977.

Sage, A. P., and White, C. C. *Optimum Systems Control*. Englewood Cliffs, NJ: Prentice Hall, 1977.

Sahlin, N. E. On second order probabilities and their applicability. Report from university of Lund, Sweden, Department of Philosophy, 1983.

Schank, R. C. *Reading and Understanding*. Hillsdale, NJ: Lawrence Erlbaum Associates, 1982.

Schank, R. C. *Conceptual Information Processing*. North-Amsterdam: Holland, 1975.

Schank, R., and Abelson, R. P. *Scripts, Plans, Goals, and Understanding*. Hillsdale, NJ: Lawrence Erlbaum Associates, Inc., 1977.

Schank, R. C., and Colby, K. M. (Eds.), *Computer Models of Thought and Language*. San Francisco: Freeman, 1973.

Schefe, P. On foundations of reasoning with uncertain facts and vague concepts. *International Journal of Man Machine Studies* 12:35–62, 1980.

Silverman, B. G. Analogy in systems management: An information processing view with implications for comparison guiding aids. Department of Engineering Administration, George Washington University, February, 1983.

Sternberg, R. J. *Intelligence, Information Processing, and Analogical Reasoning: the Componential Analysis of Human Abilities.* Hillsdale, NJ: Erlbaum, 1977.

Sternberg, R. J. (Ed.), *Handbook of Human Intelligence.* Cambridge: Cambridge Univ. Press, 1982.

Svenson, I. Process Descriptions of decision making. *Organizational Behavior and Human Performance, 23:*86–112, 1979.

Tolman, R. C. Cognitive maps in rats and men. *Psychological Review 55:*189–208, 1948.

Toulmin, S., Rieke, R. and Janik, A. *An Introduction to Reasoning.* New York: McMillan Publ., 1979.

Tversky, A., and Kahneman, D. The framing of decisions and the psychology of choice. *Science 211*(January 30):453–458, 1981.

Tweney, R. D., Doherty, M. E., and Mynatt, C. R. *On Scientific Thinking.* New York: Columbia University Press, 1981.

Ungson, G. R., and Braunstein, D. N. (Eds.), *Decisionmaking: An Interdisciplinary Inquiry.* Boston, MA. Kent Publishers, 1982.

Wason, P. C., and Johnson-Laird, P. N. *Psychology of Reasoning: Structure and Content.* Boston, MA. Batsford, 1972.

White, C. C., Sage, A. P., and Scherer, W. T. Decision support with partially identified parameters. *Large Scale Systems 3:*177–198, 1982.

Wilensky, R. Meta-planning: Representing and using knowledge about planning in problem solving and natural language understanding. *Cognitive Science 5:*197–233, 1981.

Winograd, T. *Language as a Cognitive Process.* Reading, MA: Addison Wesley, 1983.

Winograd, T. Frame representations and the declarative-Procedural Controversy. In Bobrow and Collins, 1975, pp. 185–210.

Winston, P. H. Learning and reasoning by analogy. *Communications of the ACM 23*(12):689–703, 1980.

Winston, P. H. *Artificial Intelligence.* Reading, MA: Addison-Wesley, 1977.

Winston, P. H., and Brown, R. H. (Eds.), *Artificial Intelligence: An MIT Perspective.* Cambridge, MA: MIT Press, 1979.

Winterfeldt, D. Von Structuring decision problems for decision analysis. *Acta Psychologica 45*(1–3):71–93, 1980.

Wohl, J. G. Force management requirements for Air Force Tactical Command and Control. *IEEE Transactions on Systems, Man and Cybernetics SMC-11*(9):618–639, 1981.

Zadeh, L. A. Fuzzy sets. *Information and Control 8:*338–353, 1965.

Zadeh, L. A. Fuzzy logic and approximate reasoning. *Synthese 30:*407–428, 1975a.

Zadeh, L. A. The concept of a linguistic variable and its application to approximate reasoning. *Information Sciences,* Part I, II and III, 1975b, Vol. 8, pp. 199–249, pp. 301–357 and Vol. 9, pp. 43–80.

Zadeh, L. A. Outline of a new approach to the analysis of complex systems and decision processes. *IEEE Transactions on Systems, Man, and Cybernetics SMC-3*(1):28–44, 1973.

HUMAN DECISION PROCESSES IN MILITARY COMMAND AND CONTROL

Joseph G. Wohl, Elliot E. Entin, David L. Kleinman
and Krishna Pattipati

ABSTRACT

Military command and control, while heavily supported by an ever evolving technology of sensors, computers, displays, communications and weapons, remains fundamentally a human decisionmaking activity. This chapter presents several examples of C^3 systems and describes the kinds of decisions that are made in each. A review of pertinent literature is provided, followed by a detailed description of individual decision processes in command and control, a brief review of recent work on teams of interacting decisionmakers, and a discussion of problems and approaches to decision task isolation and representation.

I. INTRODUCTION

In addition to its weapon systems, each military service acquires and deploys a wide variety of radars, computers, and communications equip-

Advances in Man-Machine Systems Research, Volume 1, pages 261–307.

ment to support its capability to defend and fight. In this regard there has
been an inordinate tendency to consider these devices as "things." But
in order to understand more clearly what we must do to design them
better, we first need to recognize that these "things" really are extensions
of the limited human capacity to sense, process and communicate quickly
and over long distances; and finally, we need to recognize that the process
of military command and control similarly is an extension by means of
systems of equipment, personnel and procedures of the human decision-
making process imbedded in a military organization.

In this chapter we discuss the process of decisionmaking in command
and control and summarize the current state of knowledge in this area.

II. ON DESCRIBING COMMAND AND CONTROL

A military command and control (C^2) organization is comprised of military
personnel (human operators) carrying out the command and control process. This process is a coordinated set of information gathering and decisionmaking functions, carried out with the objective of effective force
application, i.e., of best utilization of weapon systems in a battle environment. The C^2 process is supported by the command, control and communications (C^3) system. The C^3 system is a collection of sensor, information processing, and communications subsystems that allows the C^2
organization to receive and analyze information from and transmit information to the battle environment, and that in addition facilitates information interchange between the members of the C^2 organization. Finally,
the C^2 process along with the system and organization which support it,
all operate in support of one or more specific military missions such as
air defense, space defense, close air support of an air-land battle, or antisubmarine warfare coordination. Figure 1 indicates the interrelationships
among these entities.

Neither the military mission, the battle environment, the C^3 system,
nor the C^2 organization is particularly hard to describe, at least in principle. The battle environment contains friendly and enemy weapon systems and other objects of interest. The C^3 system consists of radars,
computers, radio transmitters and receivers and similar physical objects.
The C^2 organization is a collection of military personnel. It may be a
matter of convenience to consider the operator of an anti-aircraft artillery
(AAA) system to be part of a weapon system in one situation or as part
of the command and control system in another, but this causes no conceptual difficulties.

What is much more difficult to describe is the C^2 process. This process
consists of a collection of human activities, organized to accomplish certain military goals. Unfortunately, there exists no precise, standard tech-

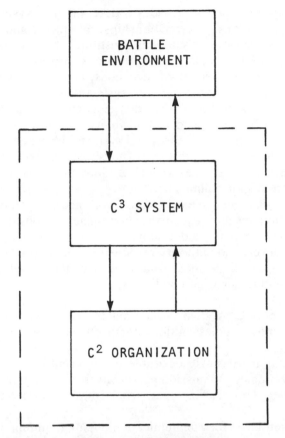

c^2/c^3 SYSTEM (c^3 SYSTEM FOR SHORT)

Figure 1. Military Command and Control.

nique for even describing the C^2 process, much less for analyzing or for designing it. The difficulty in describing this process arises because it is a dynamic process, carried out by a team of human decisionmakers and operational personnel, who may be distributed over a considerable geographic region, and who are forced to operate under conditions of considerable uncertainty in achieving both their individual goals and those of the overall organization in which they operate.

A. Perspectives

Command, control, and communications (C^3) systems can be viewed, and hence analyzed, from several perspectives, depending upon the purpose of the viewer. For example, consider the following:

- *A Systems Perspective* provides a description of system and sub-system capabilities, interrelationships, interfaces, and data flows. It answers questions about what constitutes the system.
- *A Functional Perspective* adds functions, processes, procedures, doctrine, routine (structured) decisions, and information flows. It answers questions about how the system works.
- *An Organizational Perspective* adds geographic and hierarchical distribution of authority and responsibility and flow of control, and associates functions/processes and systems/subsystems with organizational levels and authority/responsibility units (i.e., decisionmakers). It answers questions about how the system is organized and where responsibilities lie.
- *A Mission Perspective* adds elements of uncertainty (enemy versus own capabilities and objectives, environmental constraints and impact), and *unstructured* decisions, and associates them with organizational levels and authority/responsibility units (decisionmakers). It answers questions about what objectives the system is supposed to meet and under what conditions.

A description of a C^3 system is incomplete, for analytical and design engineering purposes, unless all four perspectives are appropriately represented.

It is becoming increasingly desirable to represent the physical, functional, organizational, and mission perspectives of C^3 systems in a single, unifying manner. This growing need arises from two requirements:

- A common basis for communication among requirements determiners, operational personnel, tacticians, analysts, and system designers.
- An operational (i.e., mission-oriented) basis for engineering analysis and design.

Engineering analysis and design has been successful at the subsystem level to the extent that systems requirements could be decomposed or broken down into subsystem requirements and translated into subsystem specifications. Decomposition and definition of requirements is a fairly straightforward task for well-structured, well-partitioned systems, especially those with centralized command, hierarchical connectivity and strict division of authority. However the increasing emphasis in new systems on survivability through distributed processing and control and on distributed decisionmaking is forcing the design of less well-partitioned systems with greater connectivity and redundancy, for which existing representation/decomposition techniques are not well suited. The time-honored system functional block diagram and specification family tree, while appropriate to equipment and well-partitioned-system description, cannot without severe modification be used to represent the human, organizational and procedural aspects of an adaptive, multimode system in which human decisionmaking is a critical activity. For such systems, representation/decomposition tools must be capable of representing successively lower and more detailed levels of the physical system, as before. In addition, the system's functions and processes, the structure and flow of control authority throughout the system, the goals and objectives of the system, the decisionmaking portions of the system, and the operational (i.e., mission) perspective appropriate at such levels must also be represented. We shall return to this problem in Section VII in a discussion of decision task isolation problems and approaches.

B.   Observations and Issues

The following observations describe the essential elements of military command and control. First, it consists of information processing and decisionmaking activities by military personnel. Second, in carrying out these activities, higher level human cognitive skills rather than lower level human sensory-motor skills are most relevant.

Third, the information in C^3 systems usually concerns the past, present, and future location, identity, and certain other attributes of various objects. These objects may be actual physical objects (e.g., tank, ICBM) or abstract objects consisting of a collection of physical objects (e.g., tank battalion, missile salvo), and may be under friendly, neutral, or enemy control. Information concerning these objects comes from various sensors (including human observers) and is subject to a variety of errors and time delays. The information may be presented to C^2 organization personnel by a variety of means including visual displays and auditory or written communication. These personnel must attempt to sort valid from erroneous sensor reports, and to develop a coordinated, correct assessment of the battlefield situation, for communication to appropriate decisionmakers. Since data is received sequentially in time, since the past and future as well as the present battlefield situation is of concern, and since information is highly time critical,[1] the information processing activities in C^2 are highly *dynamic* in character. Human limitations with respect to processing capacity, errors, etc. are critical in determining how well the information processing activities in the C^2 environment are carried out.

Fourth, the decisionmaking problems in C^2 generally involve the positioning and timing of friendly weapons systems with respect to enemy forces. This positioning may be explicit, as when an intercept officer directs a fighter/interceptor or when an army group commander positions an armored division, or may be implicit as when the commander of a surface-to-air-missile (SAM) battery directs a particular SAM to engage an attacking aircraft. Thus the decisionmaking problems generically involve dynamic sequencing, resource allocation, and scheduling problems, with the dynamics due to the motion of friendly and hostile forces. There are usually additional second-level decisions spawned by the scheduling/resource allocation decisions; e.g., when a target is assigned to an antiaircraft artillery (AAA) site, an engagement mode must be selected, or when a fighter-bomber is assigned to attack an interdiction target, its armament must be selected. These second-level (implementation) decisions are referred to in the following section as supervisory tasks.

Fifth, it should be emphasized that both the information-processing and decisionmaking activities of personnel in the C^2 organization are conducted under conditions of uncertainty. The information provided by sensors and human observers is incomplete and contains distortion, noise, and gross errors even in the best of circumstances. Under battle conditions, enemy decoys, countermeasures, deception, etc. serve to further complicate the already difficult decisionmaking and information processing tasks.

A sixth point that must be emphasized is that the information processing and decisionmaking activities in the C^2 organization are distributed among multiple human operators, who may be in different geographical locations. These humans must coordinate their activities in such a way as to achieve the objectives of both the C^2 organization and the current military mission. The distributed nature of the C^2 process greatly complicates its description, analysis, and design. Multiple subprocesses may be going on in parallel, the right information must be provided to the right decisionmaker at the right time, and workload must be allocated properly so that bottlenecks are avoided, to mention but a few of the difficulties associated with the distributed nature of the C^2 process.

A final point to be emphasized is the *dynamic* nature of the information processing and decisionmaking activities comprising the C^2 process. The *past* movement of an enemy organization helps one to estimate its *future* location. The decision to send a close air support aircraft or a torpedo-carrying helicopter in response to a request *now* forecloses for a finite time the option of using that platform to respond to a *future* request.

Thus we see that the basic difficulty in understanding the C^2 system is that this ''system'' encompasses the distributed, dynamic information processing and decisionmaking activities, carried out under uncertainty, of the personnel constituting the C^2 organization all in support of a given military mission. Since even understanding the performance of a single human performing a single well-defined task is a nontrivial affair, it is not surprising from whence the difficulty in understanding C^2 systems arises.

We now turn to a description of several specific examples of C^3 systems.

III. THREE EXAMPLES OF C³ SYSTEMS

There is a wide variety of C³ systems and organizations to be considered, including U.S. tactical and strategic systems, as well as foreign C³ systems that our military forces must be prepared to counter. It is necessary to be able to describe, analyze, design (in the case of U.S. systems), and design countermeasures for (in the case of enemy systems) these systems. This task is complicated by the fact that it is impossible to observe these systems in a fully stressed mode of operation except in wartime; and even peacetime doctrinal, training and exercise observations may be limited in the case of threat systems. Following are brief descriptions of examples from each of the three categories noted above.

A. U.S. Air Force Tactical C²

In U.S. Tactical Air Command and Control, the terms "command" and "control" can be used to distinguish between two separable categories of decisions: those which are made long before an aircraft leaves the ground, and those which are made when an aircraft is on "ready alert" or is already airborne. These two categories have been called by various names; we use the term planning and commitment to represent the preflight, command-oriented, broadly based decision category, and control and coordination to represent the alert/inflight, control-oriented, narrowly focused, and time-critical decision category.

1. Planning and Commitment

Figures 2 and 3 describe the four basic planning and commitment decisions made by Air Force tactical commanders at various levels. Information required for these decisions includes

- the status of enemy forces and its rate of change;
- the status of friendly forces and its rate of change;
- the environment: weather, terrain, darkness
- command guidance: policy, rules of engagement, campaign priorities

As indicated in Figure 2, the time-criticality of information required for planning and commitment decisions generally decreases with the decisionmaking level; the higher-level commander deals with planning for tomorrow's war, while the unit-level planner is concerned primarily with today's missions. On the other hand, the degree of information aggregation required generally increases with the decisionmaking level: the unit level planner needs the latest information on location, characteristics, and activity of specific enemy SAM sites on or near the planned target ingress

DECISION	COMMAND LEVEL	DEFINITION	RESULT	EXAMPLE	PERIOD
APPORTIONMENT (ALLOTMENT IN NATO)	JOINT TASK FORCE COMMANDER (AAFCE IN NATO)*	GROSS APPORTIONMENT & PRIORITIZATION BY PERCENT OF WHATEVER RESOURCES ARE AVAILABLE, AGAINST MISSIONS AND GEOGRAPHIC AREAS	PERCENT SORTIES VS. MISSION CATEGORIES	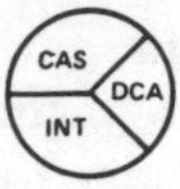 CAS DCA INT	DAYS
ALLOCATION	AIR COMPONENT COMMANDER (ATAF IN NATO)*	ALLOCATION, BY NUMBER OF SORTIES, OF AVAILABLE RESOURCES AGAINST SPECIFIC MISSIONS AND GEOGRAPHIC AREAS	NO. SORTIES VS. MISSION CATEGORIES	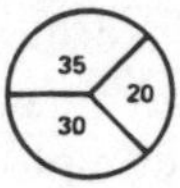 35 20 30	DAY
TASKING	TACC DUTY OFFICER (ATOC IN NATO)*	COMMITMENT OF AIRCRAFT AND ARMAMENT FROM VARIOUS SOURCES AGAINST SPECIFIC TARGET AREAS OR TARGETS	NO. SORTIES FROM EACH AIRBASE VS. EACH TARGET AREA OR TARGET		DAY/HOURS
DETAILED UNIT-LEVEL PLANNING	TACTICAL AIRBASE UNIT LEVEL	ASSIGNMENT OF SELECTED AIRCRAFT TO SELECTED MISSIONS AND TARGETS; PLANNING OF INGRESS AND EGRESS ROUTES TO MINI-MIZE ATTRITION AND MAXIMIZE EFFECTIVENESS	TACTICS, TAIL NUMBERS, AIRCREWS, AND PREFLIGHT BRIEFINGS; DETAILED FLIGHT PLAN FOR EACH SORTIE; CALL SIGN AND IFF CODE DESIGNATION; WEAPON LOADS AND FUZING; ETC.		HOURS

CAS = CLOSE AIR SUPPORT
DCA = DEFENSIVE COUNTER AIR
INT = INTERDICTION
(OTHER MISSIONS NOT SHOWN)
* SEE FIGURE 4 FOR DEFINITIONS

Figure 2. Current USAF Tactical Planning and Commitment Decisions.

route, while the higher-level commander needs only enough information about enemy SAMs to help allocate defense suppression forces.

The relationship between planning and commitment decisions and required information inputs is also shown in Figure 3. The two-headed arrows in this figure are meant to emphasize the cooperative aspects of the decision process. A decision at one level requires both inputs from and coordination with levels immediately above and below.

2. Control and Coordination

Changes may also be needed in the control and coordination of mission aircraft. Once in flight, they will be assisted and sometimes redirected by the control and coordination hierarchy of Figure 4. Flight plans will call for certain actions at critical times and locations, depending upon the mission. Interdiction, close air support, suppression of enemy defenses, offensive and defensive counter-air, and associated support missions may or may not have these actions defined in detail at takeoff. Ground targets appear and disappear as they move; support missions fail; priorities change; objectives and plans are superseded; new information is received; and a mission plan with which the pilot took off must be modified accordingly, in-flight if appropriate and possible. It is under these circumstances of high stress and severe time constraint that near-real-time de-

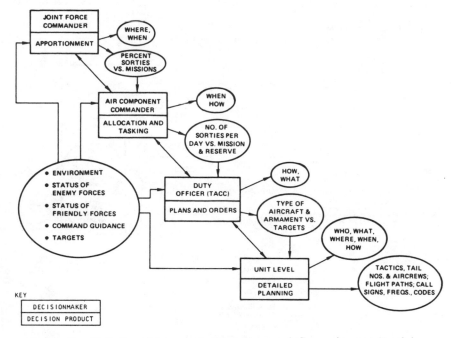

Figure 3. Relationship among Planning and Commitment Decisions.

cisions involving threat warning, rerouting, retargeting, rescheduling, and other types of dynamic reassignment and mission modification must be made. Especially critical will be the areas of aircraft identification to minimize fratricide, in-flight provision of ground target updates, and threat warning and defense suppression coordination to improve strike mission effectiveness and reduce attrition of friendly forces, all of which require effective communications.

Figure 5 summarizes the critical decisions and indicates how the time constraints and information needs vary as the decision level descends. Note that the higher level decisions take more time to execute or modify because of their strong dependence on materiel, maintenance, and personnel scheduling and movement; and arbitrary changes will have little immediate effect because of the "inertial mass" of these supporting systems. This suggests that decision-making procedures must somehow be significantly compressed (e.g., through more downward delegation and management by exception) in order to reduce both the number of levels involved in a given C^2 function and the impact of holding times and processing delays at each level. In addition, the lower level decisions require more informational detail while higher level decisions require less detail and more aggregation. The impact of such changes on C^3 system design

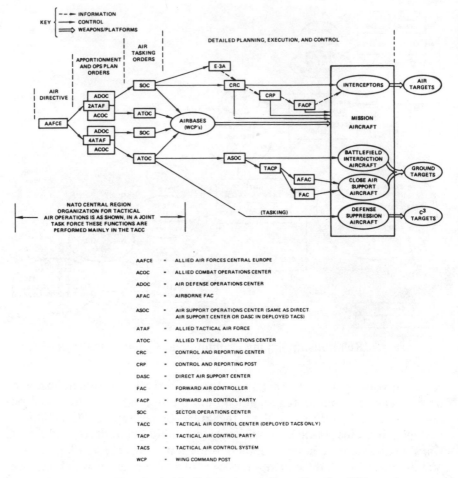

Figure 4. Tactical Air Control and Coordination Hierarchy.

requirements, especially in the area of decision support, needs to be carefully assessed.

B. U.S. Air Force Strategic C²

Of particular interest in strategic command and control are the threat warning and attack assessment (TWAA) decisions along with the relationships between them and their association with the various command levels. These are summarized in Figures 6 and 7. Note the parallelism between these figures and those representing the tactical command and control decisions and command levels in the previous section. The fundamental difference is that of time stress: when a strategic event occurs (such as the destruction of a friendly satellite or the detection of a ballistic missile raid) the total amount of time available between event occurrence and strategic weapon release may be less than 30 minutes. This means that all subsequent activities must be carefully thought out long beforehand, completely preplanned, and highly proceduralized. Essentially, the detection of such an event must serve to trigger a preplanned sequence of activities, with different event types triggering different activity sequences. This requires a rapid categorization of event type and a mapping of the event type onto appropriate procedures. The Threat Warning and Attack Assessment (TWAA) Center in NORAD is a central focus for rapid processing of event information as a primary input to the NORAD command post's function of event evaluation, and into CINCNORAD's primary function of event assessment, as well as to the National Command Authority as an input to final weapon release decision and triggering of portions of the SIOP. Here too, a decision at one level requires both inputs from and coordination with levels immediately above and below. In addition the U.S. strategic forces require continuous inputs from all levels

DOMAIN	DECISION CATEGORY	DECISION TYPE	TIME AVAILABLE FOR DECISION	DEGREE OF INFORMATION AGGREGATION REQUIRED
PREFLIGHT	PLANNING AND COMMITMENT	APPORTIONMENT ALLOCATION TASKING DETAILED PLANNING	(MORE TIME) ↑	(MORE INFORMATION LESS DETAIL) ↑
INFLIGHT	CONTROL AND COORDINATION	RESOURCE ALLOCATION TARGET REASSIGNMENT TARGET/UPDATE INTERCEPT CONTROL THREAT WARNING	↓ (LESS TIME)	↓ (LESS INFORMATION MORE DETAIL)

Figure 5. Summary of Critical Tactical Command and Control Force Management Decisions.

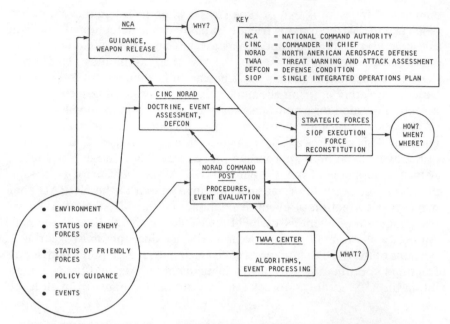

Figure 6.   Relationship among Strategic Threat Warning and Attack
Assessment (TWAA) Decisions.

in parallel, in order to eliminate the nodal delays which would otherwise
occur in information transmission.

As demonstrated at NORAD during the summer of 1980, a fundamental
problem of threat warning and attack assessment is the problem of false
alarms. A related problem, one which has not yet been demonstrated but
which will inevitably arise, is that of uncertainty in the input information.
At present, this is dealt with on the fairly low-level basis of establishing
confidence levels in sensor input data. However, a complex set of events
could result in a level of uncertainty as to enemy action, enemy capability,
and/or enemy intent which could render the assessment/release process
ineffective if that complex set of events is properly planned by a clever
enemy. Thus, the question of event-pattern uncertainty and its reduction
will take on increasing importance for TWAA in the future.

Figure 7 summarizes some of the critical C^2 decisions in TWAA and
indicates how the time constraints and information needs vary as the
decision level descends. Note that those at the higher level take more
time to execute or modify because of their strong dependence on logical
analysis and value judgements. In addition, the lower level decisions require more informational detail regarding the nature of the set of events

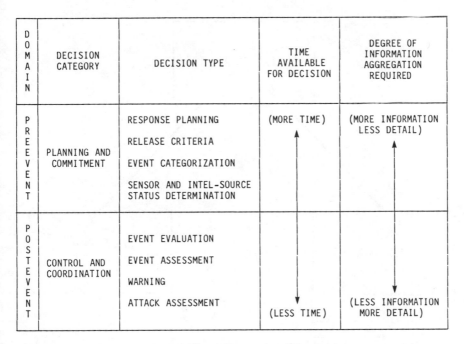

DOMAIN	DECISION CATEGORY	DECISION TYPE	TIME AVAILABLE FOR DECISION	DEGREE OF INFORMATION AGGREGATION REQUIRED
PREEVENT	PLANNING AND COMMITMENT	RESPONSE PLANNING RELEASE CRITERIA EVENT CATEGORIZATION SENSOR AND INTEL-SOURCE STATUS DETERMINATION	(MORE TIME) ↑	(MORE INFORMATION LESS DETAIL) ↑
POSTEVENT	CONTROL AND COORDINATION	EVENT EVALUATION EVENT ASSESSMENT WARNING ATTACK ASSESSMENT	↓ (LESS TIME)	↓ (LESS INFORMATION MORE DETAIL)

Figure 7.   Summary of Critical Strategic TWAA-Related Decisions.

and/or presumed attack while the higher level decisions require less detail and more aggregation.

It is clear that the growing complexity of strategic-associated event analysis and decisionmaking in a high-false-alarm-rate environment, when coupled with increased data input resulting from improved sensor coverage and sensitivity and reduced time available for decision (e.g., due to the potential for fractional-orbital attack), can result in high payoff opportunities for C^2 systems analysis and design.

C.   Enemy Air Defense C^2

Military commands of potential belligerent countries are subdivided into hierarchies, as are U.S. military commands. Figure 8 depicts a generic threat command structure. The tactical air force portion of this figure has a command and control system similar in certain respects but dissimilar in other respects to the corresponding USAF C^3 system discussed above. The C^3 system of potential enemy ground forces is very much of concern to U.S. planners, as a substantial portion of an enemy air defense capability will be under the control of those ground forces.

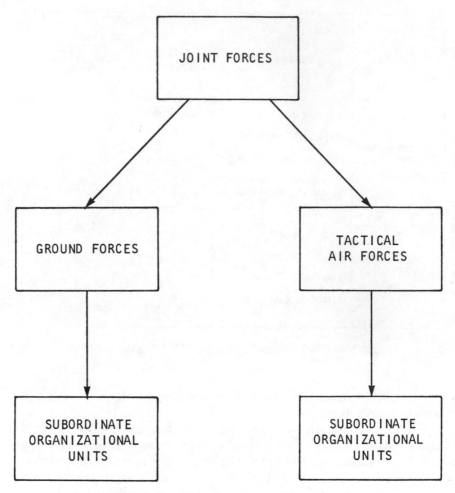

Figure 8.   Threat Command Structure.

Figure 9 illustrates in a schematic way the components of an enemy air defense system and its environment. Roughly, the system is driven by targets and jammers that are external to it. Its sensors detect, track, and identify targets. Information from the sensors is distributed to command and control centers that evaluate the air picture and assign weapon systems to engage targets. Information must be transferred to the weapon systems to permit target acquisition. All of this would have to be accomplished in the face of continuous jamming as well as physical attack by attacking forces.

Command and control decisions must be made at all organizational levels associated with the threat air defense system. At the lowest level of the threat ground forces organizational structure are highly mobile air defense assets: vehicle mounted surface-to-air missiles (SAMs), antiaircraft artillery (AAA), shoulder fired missiles, and machine guns. The higher levels may involve warning, direction and coordination systems consisting of sensors (radar and radio intercept equipment), communication links, computers, and personnel, as well as longer range but somewhat less mobile SAMs.

The air force component of an enemy air defense C^3 system consists of individual fighter/interceptors (which are themselves extremely complex man-machine systems) at the lowest organizational level, with a complex collection of ground control intercept (GCI) centers, and filter and fusion centers at higher organizational centers. A variety of sensor and communication systems are used at each level of the organizations.

At all levels, the key analytical problems of concern to the U.S. forces are those of minimizing inferential uncertainty: Given a set of data about an enemy C^3 system which may include, for example, equipment characteristics, operating instructions, employment doctrine, organizational hierarchy and connectivity information and so on, how does one go about inferring the specific data flows, timelines and procedures necessary to fully understand how selected C^2 functions are carried out? Clearly, an understanding of enemy air defense C^2 is essential to the U.S. planners

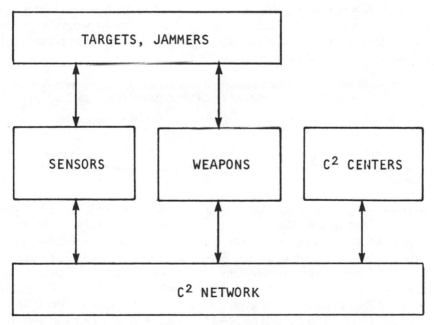

Figure 9.   Components of Threat Air Defense Network.

so that engagement analyses can be carried out, penetration route planning can be accomplished, and countermeasures can be designed. To achieve such an understanding, the information processing and decision-making processes of enemy commanders must be characterized.

IV.   REVIEW OF LITERATURE ON C² DECISIONMAKING

In the preceding discussion, we have described three examples of C³ systems and have indicated some of the generic issues associated with their analysis and design. We will now briefly review some of the literature concerning these issues. Our review will generally be limited to recent and open literature focused on C² decisionmaking as opposed to surveillance, fusion, or communication; to papers and reports concerning discussion of generic issues as opposed to specific systems; and to C²-specific contributions as opposed to general work on decisionmaking and information processing, which has been reviewed in depth (Sage, 1981). We will see that a major weakness of existing work is that the key roles and limitations of the humans who operate the C³ systems are not adequately considered or represented.

A.   Theory and Assessment of C²

It is widely recognized that there does not presently exist a satisfactory C² theory, and that the lack of theory is a serious impediment to the satisfactory design and procurement of C³ systems. Several recent quotations bear out this claim:

> The challenge we face as a community is to develop an ability to quantitatively assess the utility of our command and control systems (MITRE, 1980).

> . . . the [C³] hardware technology provides us greater promise than it has rewards . . . (Van Trees, 1980).

> . . . neither the Phase I investigation nor current related efforts apart from CONSTANT QUEST have substantiated a practical quantification capability (Welch, 1981).

In response, a variety of educational and research efforts have been initiated. These include the new joint-services C² curriculum at the Naval Post Graduate School; the research programs at MIT sponsored by the Office of Naval Research and the Air Force Office of Scientific Research; the MIT/ONR workshops (Athans, 1978; Athans, 1979; Athans, et al., 1980; Athans, et al., 1981); a series of workshops under Air Force and Office of the Secretary of Defense sponsorship (Thrall, et al., 1976; Wohl, 1980a; MITRE, 1980); the 1980 NATO conference on command and control (Wohl, 1980); and the special sessions of the recent IEEE Conference on Decision and Control (Athans, 1980). The Proceedings of the various workshops in particular provide a valuable indication of the state-of-the-art in C^2 modeling, analysis, and design.

The C^2 literature surveyed can be placed into four broad categories:

- qualitative discussions of C^2 issues,
- advocacy of a particular academic discipline or disciplines as the basis for a C^2 theory,
- assessment methodologies for C^2, and
- decision aids for C^2.

Of course, many of the papers overlap some (or even all) of these categories. In general, only the fourth category of work explicitly treats the human decisionmaker.

Most extensive are the qualitative papers. While a detailed discussion of individual papers will not be attempted here since there are so many of them, some of the more significant conclusions of these papers are:

- There is at present no adequate theory for C^3 system analysis and design.
- The basic difficulty is that the hardware and software elements of a C^3 system serve only to facilitate the information management and decisionmaking processes of commanders, and these processes are poorly understood.
- C^3 system theory must predict measures of overall system utility or effectiveness (e.g., attrition rates, FEBA movement, mission success) as opposed to measures of individual subsystem performance (e.g., storage capacity, access time, bit rate, signal-to-jammer margin).
- A C^3 system theory is needed to provide a framework to integrate analytic findings, research results, empirical data, and judgement.
- Analytical models of C^3 systems must be validated by comparison of predictions with detailed simulations, man-in-the-loop test bed simulations, war games, and full-scale exercises.
- A C^3 theory has potential applications in aiding front end investment decisions, in facilitating the design process, in training, and in devising counter-C^3 strategies.

Various academic disciplines have been put forward as the basis for a theory of C^3, although the bulk of the analyses to date have had a heavy military operations research flavor. Due to the dynamic and stochastic nature of the C^3 problem, various investigators have suggested that control theoretic methods be employed to model C^3 systems. Cybernetic feedback models have been suggested by some (Van Trees, 1980; Lawson, 1980a; Lawson, 1980b; Rona, 1980; Zracket, 1980; Athans, 1980); while others have emphasized that distributed rather than classical, single feedback loop control theory is the correct setting for those models (Sandell, 1980; Tenney, 1980).

The key role of human decisionmakers in C^3 systems naturally leads to the consideration of decision analysis and related methodologies in cognitive science. Modrick (1976) and Wohl (1979, 1981) have emphasized the issue of decisionmaking style in tailoring a C^3 system to a particular decisionmaker, and categorizing and choosing decision aids. Many authors have advocated (and presented results obtained from) man-in-the-loop testbed simulations as a means of development and validating C^3 concepts. This includes facilities of military research and development establishments such as the Army Research Institute, Air Force Aerospace Medical Research Laboratory, and Naval Ocean Systems Center, as well as various training and operational facilities.

Artificial Intelligence (AI) techniques have also been advocated as a means of modeling C^3 systems. A model comprised of a collection of interacting knowledge-based systems was proposed by Aldrich (1980). Certain artificial intelligence-based techniques (production rule systems and signature tables) had an impact on the Air Force Headquarters Studies and Analysis TAC ASSESSOR simulation model [Leedom, 1980]. A limitation of the AI techniques proposed to date is that they depend on interviews with experts to develop a representation of their decisionmaking processes. Thus the techniques would seem to be limited to the description of existing systems for which experts presumably are available; the way in which AI techniques can be used to assist in synthesizing new structures is less obvious.

Papers discussing specific quantitative assessment methodologies for C^3 systems are not so numerous, and will be discussed at greater length.

Lawson (1978, 1980a, 1980b) has described an appealing "thermodynamic" model of a C^3 system which relates the military pressure a commander can exert over a volume of space to his weapons system and C^3 system assets. This formulation is closely related to the classical Lanchester equations of warfare, since these can be derived from Lawson's model if it is assumed that attrition during an engagement is proportional to military pressure. Using his formulation, Lawson is able to explicitly compute a force multiplier in terms of the reduced number of combat air patrol aircraft (equivalently, increased effectiveness/aircraft) resulting from a reduction in C^3 system decision delay.

The key ideas in Lawson's calculations are to take a conventional engagement analysis and to vary the parameters of that analysis in accordance with the presumed effect of an improved C^3 system.

Everett (1981) has done the same thing using Lanchester's equations directly and has demonstrated the effects not only of decision delays but also of increased sensor accuracy and resolution. Other approaches of this type are described in the literature, albeit with more elaborate, simulation-based analyses. Miller (1980a, 1980b) describes an analysis of the C^3 issues associated with the defense of a carrier task force from cruise missile attack. An improved C^3 system ("high coordination") is modeled as inducing a reduction in the probabilities of fratricide and the probabilities of assigning multiple or no interceptors to a target. Alberts (1980) describes a study in which a new communications capability is assumed to increase the ability to quickly locate targets and bring them under fire, and this capability is reflected in an increase in certain of the attrition coefficients in a Lanchester-type engagement model. A similar approach is described by Leedom (1980), in which a particular C^3 enhancement is presumed to result in an increase in close air support aircraft sortie rates.

A limitation of the preceding approaches to quantitative C^3 system assessment is that the C^3 system elements and the C^2 decision process of human commanders they support are not modeled explicitly. We shall now describe two examples of explicit modeling of the C^2 decision process.

B. Computer Simulation: TAC ASSESSOR Model

This simulation model (Leedom, 1980; Welch, 1980) is a very large and complex air/ground combat simulation that includes an explicit representation of the C^2 organizations/decisionmakers (down to the pilot level) involved in tactical air reconnaissance and attack missions. It is noteworthy since the effect of misperceptions and delays in information can be evaluated with respect to their effect on the battle outcome. It uses AI techniques (in a general sense), e.g., planning is performed using a heuristic tree search approach and situation assessment is loosely based on frame theory.

The TAC ASSESSOR model is a significant accomplishment, but is not without its shortcomings. First, the model is a highly detailed simulation with extremely long run times.[2] Second, the model does not incorporate (at least explicitly) the limitations on human decisionmaking ability that have been described in the cognitive science literature over the past 20 years. Third, the model does not incorporate the effects, or alternatively, specify the requirements for, communication and computational resources. Finally, the model is a detailed representation of an existing system; it is hard to see how the methodology would apply to the conceptual design of a new system in which there were not "expert" decisionmakers already existing. None of these remarks is intended to be a criticism of the TAC ASSESSOR model per se, which is well suited to the purpose for which it was intended. The point is simply that the model has limitations as a basis for general C^3 systems analysis and design methodology, and for the study of human performance in C^3 systems in particular.

C.   Mathematical Models: The Commander's "Mental Model"

Wohl (1981a) developed an extended form of Lanchester's equations, encompassing the effects of second-echelon enemy ground forces, friendly close air support and interdiction air resources, and FEBA movements, to represent the air component commander's "mental model" of a tactical air/ground engagement situation. (See Section V for further detailed discussion of this concept.) The resulting equation is then solved for minimum friendly attrition subject to constraints on FEBA movement and initial resources. In addition, Wohl shows how a stochastic version of this model can be used to extract a new generic measure of C^3 system effectiveness, namely, rate of reduction of uncertainty. This approach, while still highly theoretical in nature, nonetheless provides a beginning basis for both formal analysis and experimental verification.

In a related effort, Lauer, et al., used a Lanchester model similar to Wohl's to describe the interactions among Air Force commander's decisionmaking processes at several levels of the Tactical Air Force command hierarchy (Lauer et al., 1980). The approach is to develop normative models of the individual commanders, and to use the domular approach of Tenney and Sandell (1981a, 1981b) to describe the coupling between the individual normative models.

D.   Decision Aiding

The final segment of the C^3-specific literature that we will discuss is decision aiding. Basically, there are two types of decision aids: those based on computer science and artificial intelligence (AI) techniques and those based on decision theoretic techniques. One example of an AI-based decision aid is LADDER, which permits natural language queries (e.g., where is the aircraft carrier ENTERPRISE?) of online Naval data bases. LADDER and other related techniques are being evaluated at the Naval Ocean Systems Center's Advanced C^2 Architectural Testbed facility (Brandenburg, 1980). Another AI example is KNOBS, a tactical air mission planning system (Engelman et al., 1979). Other examples of computer-based techniques are summarized by Wohl (1981b).

The literature on decision aiding based on decision theoretic techniques is primarily focused on the very simple expected utility paradigm, in which possible decisions and uncertain events are enumerated via a decision tree. The basic issues are:

- estimation of the subjective probabilities of the events and
- evaluation of the decisionmaker's utility function for the outcomes.

Examples of topics discussed in this literature are the use of graphics to facilitate decision aid utilization (Kelley, 1980), techniques (including displays) for obtaining a consensus among a group of decisionmakers on the values of subjective probability functions and utility functions (Johnston, 1980), and online evaluation of utility functions (Freedy et al., 1980).

An exception to the focus on simple normative decision tree models is the JUDGE model developed by Edwards and his colleagues at Rand Corporation in the mid-60s (Miller et al., 1967, 1968). JUDGE involves a highly sophisticated dynamic, stochastic, optimization-based decision aid for evaluating requests for close air support (CAS). This decision aid was extensively validated by a series of experiments in which the performance of human subjects aided by JUDGE and performance without it were compared. Generally, subjects with the decision aid were able to outperform unaided subjects by better than a 2 to 1 margin with respect to the utility function that measured their performance. The reason for this large increase in performance was the dynamics of the decision task: a decision to respond to a CAS request with an aircraft forecloses the option to respond to other requests with that aircraft for some future period. The "hedging" calculations needed to behave optimally in such problems are very difficult for an unaided human.

E.   Conclusions

Our review of the literature associated with command and control suggests that although there are isolated useful results, there is nothing yet available remotely approaching a general methodology specifically and completely addressing human performance in C^3 systems. The key issues appear to be, from a C^2 perspective:

- the need to explicitly represent the hypothesis and option generation and evaluation processes of individual commanders;
- the need to explicitly represent the distributed nature (and hence the coordination aspects) of C^2 information processing and decision tasks of humans;
- the need to relate these processes to the requirements for data bases, communication links, decision aids, and other hardware and software C^3 system elements; and

- the need to measure the quality of the C^2 decision process in terms of increased force effectiveness, survivability, and other measures of system *utility*.

In the remaining sections, we will elaborate on the requisite methodology to address these issues.

V. HUMAN DECISION PROCESSES IN COMMAND AND CONTROL

C^3 decision tasks have certain well-defined properties or features. The SHOR paradigm (Wohl, 1981b) provides a useful mechanism for the description of these salient features of the decision tasks of an individual decisionmaker. In essence, the Stimulus-Hypothesis-Option-Response (SHOR) paradigm of Figures 10 and 11 derived from the classic behavioral psychological stimulus-organism-response (SOR) principle (Wason, 1974). It was devised to deal explicitly with two realms of uncertainty in the decisionmaking process:

- Information input uncertainty, which creates the need for hypothesis generation and evaluation.
- Consequence-of-action uncertainty, which creates the need for option generation and evaluation.

SHOR is not meant to be a physiological or "microscopic" model; processes such as long-term memory, short-term memory, learning, etc., are considered as primitives and taken as givens. According to this view, humans interpret data received from their environment (which contains antagonists, protagonists, and "nature") in accordance with hypotheses which they formulate to represent perception alternatives. Based on these hypotheses, response alternatives are examined in order to select an appropriate response; a different response alternative set may be associated with each hypothesis. The selected response is acted upon. It influences the decisionmaker's environment and new stimuli result causing the process to repeat.

A.   SHOR Elaborated

Figure 12 is an elaboration of the SHOR paradigm to include the concept of the commander's mental model. Besides providing the decisionmaker with an internal representation of the problem, a mental model functions as a theory or framework from which to generate hypotheses. We assert that a hypothesis set derives directly from the interaction of input infor-

GENERIC ELEMENTS	FUNCTIONS REQUIRED	INFORMATION PROCESSED	
STIMULUS (DATA) S	GATHER/DETECT	CAPABILITIES, DOCTRINE; POSITION, VELOCITY, TYPE; MASS, MOMENTUM, INERTIA; RELEVANCE AND TRUSTWORTHINESS OF DATA	
	FILTER/CORRELATE		
	AGGREGATE/DISPLAY		
	STORE/RECALL		
HYPOTHESIS (PERCEPTION ALTERNATIVES) H	CREATE	C O M M A N D E R' S C A T E C H I S M	WHERE AM I?
			WHERE IS THE ENEMY?
	EVALUATE		WHAT IS HE DOING?
			HOW CAN I THWART HIM?
			HOW CAN I DO HIM IN?
	SELECT		AM I IN BALANCE?
			HOW LONG WILL IT TAKE ME TO . . . ?
OPTION (RESPONSE ALTERNATIVES) O	CREATE		HOW LONG WILL IT TAKE HIM TO . . . ?
			HOW WILL IT LOOK IN . . . HOURS?
	EVALUATE		WHAT IS THE MOST IMPORTANT THING TO DO RIGHT NOW?
	SELECT		HOW DO I GET IT DONE?
RESPONSE (ACTION) R	PLAN	THE AIR TASKING ORDER:	
		WHO WHAT WHEN WHERE HOW HOW MUCH	
	ORGANIZE		
	EXECUTE	THE NEAR-REAL-TIME MODIFICATION/UPDATE	

Figure 10. Anatomy of Tactical Decision Process—the SHOR Paradigm.

mation with a commander's mental model or internalized representation of the problem situation. If mental models act like simplified normative (predictive) theories for a decisionmaker, then hypotheses may be described in terms of (1) a set of elements or variables; (2) a set of functional relationships among the elements or variables, and (3) a set of constraints on the functional relationships.

Another key aspect of the elaborated SHOR paradigm is the introduc-

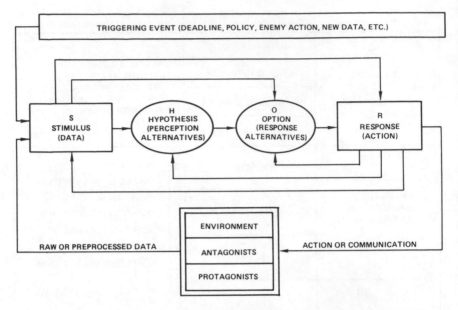

Figure 11.   Dynamics of Tactical Decision Process—the SHOR
Paradigm.

tion of the psychological concepts of attitude and cognitive style. Attitudes impact on desired outcome and indirectly on option availability and selection, while cognitive styles act through differential weights which impact on both the hypotheses and option portion of the paradigm.

The elaborated SHOR paradigm provides a means for decision task description in general terms. What is necessary in order to provide analytic substance to the SHOR paradigm is an overlay that unmasks the operator-centered elements that are amenable to analysis: data processing, information processing, hypothesis/option generation, and action selection. Figure 13 shows the relationships between the SHOR paradigm and the three[3] primary human decisionmaking elements or subtasks that we elaborate further in the following sections.

It is important to distinguish between human data processing and human information processing tasks because they involve fundamentally different activities and invoke different human limitations and capabilities. The distinction is most clearly drawn in terms of function: Data processing serves to organize data for rapid access and manipulation, while information processing serves to extract meaning from data. While this distinction will be maintained throughout the following discussions, it is clear that a high degree of interaction between them not only exists but is necessary for efficient processing.

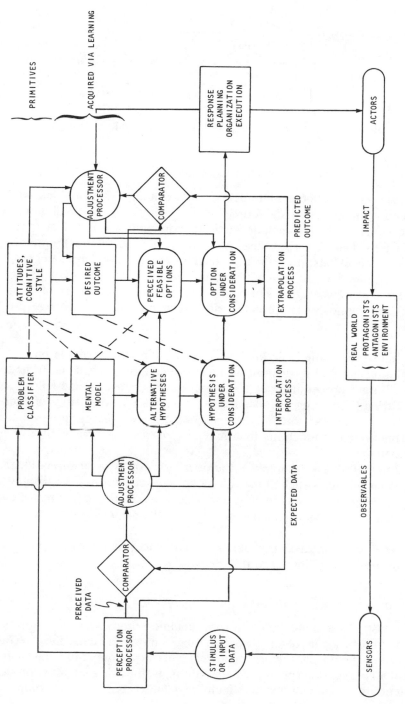

Figure 12. The SHOR Paradigm Elaborated.

285

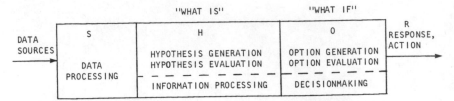

Figure 13.   SHOR Paradigm in Terms of Task Elements.

B. Data Processing Tasks

These tasks describe the interactions between the human and his available data sources. The data sources may be multimodal as well as multidimensional and may include auditory communication, visual inputs (e.g., CRTs, map displays), discrete event indicators (alarms), static inputs (e.g., checklists, procedures), etc. This (potentially) large, and often uncertain, data base is the input to the operator's data processing task.

There are three categories of data processing that support specific data processing subtasks. These categories are item, string and image processing.

- Item Processing is defined here as including those simple manipulations normally performed on a single datum or signal and basic to all other processing, such as store, update, retrieve, and so on.
- String Processing includes logical manipulations on strings of characters, such as sort, merge and compare, i.e., the usual data base management functions that generally result in a tabular display of data.
- Image Processing involves the operations on two- or three-dimensional displays such as maps, reconnaissance photos, CRT displays, etc. The operations include such things as overlaying, contrast enhancement, color enhancement, reversals, magnification, etc.

Some of the (sub)tasks that are specific to data processing include, but are not necessarily limited to the following.

1. Data Reduction, Compression and Aggregation

The task of reducing data to a form commensurate with human information processing involves the eliminating of unwanted data, the filtering of corrupting data and the organizing/categorizing of the remaining useful data. Of course, determining which data is useful is highly situation dependent, as are categories into which the data is classified. The compression/aggregation of the useful data involves its transformation into a form that reduces the (human's) fractional capacity necessary to consider it.

2. Data Search and Scan

These subtasks are associated most often with human interaction with a large data base that may be common to many users. The process of query using forms of string processing that include question-and-answer dialogues, menu selection, command languages, etc., aims at retrieving specific items from a data base. On the other hand, the battlefield itself is dynamic which necessitates repetitive scanning of the data base; thus "search" refers to what data to seek while "scan" refers to how often to sample the data. The rate at which data should be sampled is a function of the frequency of occurrence of changes in the data. For example, an operator may choose to apply a high sampling rate of photo reconnaissance over a highly dynamic region of a battlefield to reassure himself that enemy movements he has been seeing are continuing.

3. Attentional Allocation

A primary issue associated with data processing is the attention allocation among and within the various data types (modalities), i.e., inter- and intra-type attention allocations. The object of the attentional allocation task is to maximize (through judicious sequencing of data source monitoring), the information content gleaned from the data base. Attentional allocation is driven both by the data itself and by a conceptualization of what one expects or needs from the data. Conceptualization takes into consideration the purpose to which the eventual information will be put.

4. Event Detection, Recognition and Confirmation

The task of event detection is concerned with recognizing that a problem exists, or that something is happening (e.g., a missile launch or change in enemy objective). An event is detected when there exists a discrepancy between the operator's estimated (or forecasted) state of affairs, and the observed state. Thus, an event detection will typically result from receipt of new data and its correlation with current data. The detection and recognition of an event is often difficult as it may involve estimating subtle differences between two sets of highly uncertain variables: human estimates and real-world measurements. The objective of the detection task is to *correctly* recognize the event occurrence as *quickly* as possible. Clearly, this presents the operator with the trade-off among late or missed detection, confirmation, and false alarm. This trichotomy spawns what is perhaps the most fertile area for aiding the human. For example, performance improvements may be obtained by use of better alarms and alerting displays; by simplifying the format in displays; by providing relational displays which explicitly present the observed data in terms of simplified relationships; or by providing predictive displays that aid the operator to anticipate future events.

C.   Information Processing Tasks

The information processing elements of an operator task are associated with functions of data correlation, hypothesis generation and evaluation, status assessment, etc. In general terms these functions are passive, not requiring a response from the operator to the external environment, and deal largely with answering the question of "what is?" With reference to Figure 14, typical information processing subtasks include the following.

1.   Data Correlation

Data associated with a target, battlefield event, etc., will often be reported independently via a multiplicity of sensors each differing in coverage area, spectrum, resolution, response time and observables sensed. The task of sensor correlation is necessary to determine, for example, whether target data, which has been reported by different sources represents the same or different target(s). In general, data correlation involves relational processing applied to input streams. The relations themselves are outgrowths of the internal model that the processor (human or computer) holds of the engagement scenario.

2.   State and Parameter Estimation

This subtask provides continual estimation of the dynamical variables (states) that are pertinent to the military engagement. Typical of states that change continuously with time include troop strengths, positions, movements, platform locations, etc. States can also change discretely with time, e.g., perceived enemy objective, troop composition, airfield location, with large or small time interval. Variables that are constant, or change much more slowly in time than states, are defined as parameters. These include aircraft capabilities, missile ranges, radar frequencies, etc. States are generally correlated with each other, so that good estimation of certain states are likely to yield good estimates of other states, provided the human is cognizant of the underlying correlations. Understandably then, achieving the objective of state estimation with minimal uncertainty, will depend strongly on the quality of the information base and the operator's perception of the engagement dynamics.

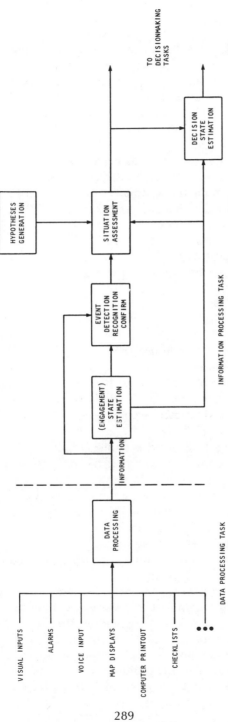

Figure 14. A Taxonomy of Display and Information-Processing Tasks.

3. Hypothesis Generation

This subtask is perhaps the most critical in a complex information-processing task. Hypothesis generation addresses the problem of how people generate a reasonable set of hypotheses (situation perceptions) and modify the set when the need arises. It involves what is commonly termed "creative thinking." Ideally, the hypotheses under consideration should be mutually exclusive and exhaustive. This is a difficult requirement and has been a major stumbling block in the application of engineering methods to describe human performance in complex organizational systems. It may not, however, be a severe problem in the engineering of systems if the operators are assumed to be well-trained and if they are provided with a knowledge base (e.g., via training) to infer possible situations causing an event.

In spite of its importance, until recently very little attention has been devoted to the issue of hypothesis generation (see [Bruner et al., 1956; Wason, 1974] for some laboratory experiments on "concept attainment" and "discover the rule" type of tasks).

Whereas information theory sidesteps the issue of meaning and deals only with symbol transmission rates and errors, understanding and describing the C^2 process requires the consideration and operational definition of meaning. We here assert that the fundamental purpose of hypothesis generation is to extract meaning from data. This definition then gives rise to a measure of performance for the hypothesis generation subtask. Since data takes on meaning in the form of hypotheses, the extraction of meaning from data can be measured in terms of hypothesis uncertainty and its rate of change. The task of hypothesis generation and testing thus is really a task of uncertainty reduction.

How human decisionmakers accomplish hypothesis generation is a critical issue, since the design of information processing and decision aids must assist them in "natural" ways. Recent work on analogical reasoning and mental models has shed some light on the process of human hypothesis generation. The analogical reasoning view asserts that when faced with a problem or a decision situation we search memory for similar or analog situations; the solutions to past problems (or minor variations of them) become the current hypothesis set to be tested (Klein, 1981). A process similar to the analog process is to scan memory for a parallel situation and then to manipulate the variables that seem to have led to solutions in the past for possible applications to the present problem.

The mental model position contends that the hypothesis set derives directly from the interaction of input information with one's mental model or internal representation of the problem situations. If mental models can act as theories, then hypotheses can be derived from them in a similar manner to that in which hypotheses are derived from theories in science; namely from the set of elements or variables comprising the theory, the functional relationships among the elements or variables, and the constraints in the functional relationships (Wohl, 1981a).

4. Situation/Threat Assessment

The task of situation assessment involves identifying the probable situation causing the observed event(s). In general terms it is the categorization by probability over a set of alternative hypotheses; i.e., pattern recognition. The inputs to this task include the event detections, the state estimates and a set of generated hypotheses for evaluation. The outputs of the task are typically the conditional probabilities of the various hypotheses being considered. The task objective is to maximize the probability associated with the (most) correct hypothesis, i.e., diagnostic accuracy as well as timeliness are the dominant concerns. There is a large literature on how *should* and how *does* the human evaluate hypotheses (Pattipati and Kleinman, 1979). The results show that the potential for human error is high due to (1) the underlying input or data uncertainties and (2) inherent human deficiencies such as conservatism, representativeness, availability, adjustment and anchoring (Kahneman and Tversky, 1972; Tversky and Kahneman, 1974; Kahneman et al., 1975).

5. Decision State Estimation

Given the assessed situation and the engagement state estimates, this task generates estimates of the pertinent variables for subsequent decisionmaking. Thus, it is this task that interfaces between information processing and decisionmaking. The decision states are transformations and compressions of the states of the engagement. Clearly, not all engagement information is necessary for decisionmaking. However, typical of decision states are the variables important for dynamic decisionmaking such as the time available for action, the time required to implement a given option, the amounts of remaining resources and their locations, etc. As with all tasks of state estimation, the objective is to provide estimates of required variables with minimal uncertainty.

D. Decisionmaking Tasks

The decisionmaking elements of a commander's activities involve planning, option evaluation, action selection and sequencing, coordination, resource allocation, target assignment, mode selection, advising, disseminating, etc. Basically, these functions are active, requiring an overt response from the decisionmaker.

Decisionmaking answers the basic question: "Given the situation, what can and will be done about it?" With reference to Figure 15, the formulation of options (action alternatives) and their evaluation prior to implementation is the canonical basis for any decision task. The major informational determinants that affect human decisionmaking behavior for a given situational assessment are:

- *Risks*—Information about potential losses and gains.
- *Alternatives*—Information about availability of and/or potential new alternatives.
- *Time*—Decision state information about deadline pressures and/or time available for deliberation.

The task of option generation, i.e., formulating the set of possible responses/actions to a situation, is functionally similar to the task of hypothesis generation and also involves "creative thinking." However, options are very often constrained or dictated in advance by higher authority (objectives), by available time and resources, by the external environment (e.g., terrain, weather), by military thought, doctrine and strategy, by rules of engagement, and so forth. Furthermore, since several different options may accomplish satisfactory outcomes, and since actions can be changed dynamically to meet evolving situations, the impact of "creative thinking" upon the C^2 process is generally greatest when focused not on option generation but on the hypothesis generation subtask of information processing.

Options are evaluated on the basis of the utilities for the potential outcomes. The utilities, which are typically multidimensional and data dependent, reflect subgoals consistent with the overall system objectives (e.g., protect people, protect facilities, counter threats, etc.). The outcome probabilities for a given option are highly dependent on the decision state estimates, which include the time and resource factors that both constrain and add urgency to action alternatives. If outcome probabilities are highly uncertain, the decisionmaker may decide to defer action in order to process/seek additional information to improve his forecasts.

The outcome probabilities, weighted by the decisionmaker's (often subjective) utility assessments, provide the cost, "subjectively expected utility" (SEU) or "expected net gain" (ENG) of an option candidate. The utilities themselves derive directly from the decomposition of the military objectives. The selection phase would consider options for implementation in the order of their ENG. If no option can be accepted, then some new action alternatives may have to be developed; failing this, the local (and sometimes higher) objectives may have to be modified. Typical decisionmaking tasks highly pertinent to human actions in C^2 applications are the following.

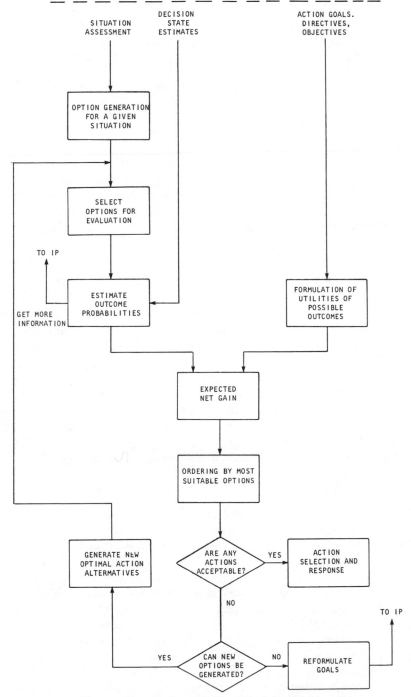

Figure 15. A Taxonomy for the Decisionmaking Process.

293

1. Supervisory "Control"

This is the simplest form of decisionmaking task beyong manual control. The task is generally associated with human actions appropriate to performing a single job in a single system, e.g., aircraft flight path generation, AAA mode selection versus single target, etc. The decisions involve the setting of parameters, switches, and modes that influence/drive the functioning of lower level—often automated—subsystems. The task is to manipulate (supervise) the subsystems by selecting alternatives from among a well-defined set of limited procedures. The objective is expressed via optimizing a metric on system performance, e.g., minimize time to target, maximize node effectiveness, etc. When there are only a few well-defined options, supervisory control simplifies to a pointwise matrix (rule-based) mapping between current situation assessment and procedure selection.

2. Sequencing

A sequencing task arises when an operator must serially perform multiple concomitant jobs. The jobs may vary in number, with each job having its own characteristics, opportunity window (deadline), and work-time requirement. The task objective is generally to complete as many jobs as possible, or complete a certain subset of jobs, etc. Sequencing is an extremely common decision task within a C^2 process. Aircraft engagement at a single AAA or SAM node, target assignments to interdiction aircraft, are examples of sequencing tasks. The options for evaluation are the different orderings associated with job selection. Utilities are associated with completed jobs (e.g., successful kills), and missed/uncompleted jobs. We note that the task of picking the procedures most likely to successfully complete an already selected job is a supervisory control task.

3. Resource Allocation

The resources available to a commander (troops, aircraft, ordnance, countermeasures, materiel, etc.) are finite. This fact has a major impact on the success probability of a selected action. On the other hand, certain actions may demand a disproportionate share of resources relative to the expected net gain of the action. Viewed in this manner, resource allocation presents a constraint that not only modifies outcome probabilities, but also introduces additional utility factors. For example, the decisionmaker must now consider the issues of resource replenishment, resource depletion, short- versus long-term needs, etc. We tend to think of resource allocation as overlaying/constraining engagement oriented scenarios, but resource allocation in terms of a planning decision can maximize the effectiveness of a fighting unit with respect to possible future situations.

4. Scheduling

The task of scheduling involves the coordination in time of two or more jobs that are being done in parallel, along with the coordination of the requirements of the jobs on the available resource pool. For example the correct time sequencing of countermeasures, strike and reconnaisance activities over a target area involves the proper coordination of a strike force and is often a critical factor in the mission's success, since lack of coordination can leave some units without air cover, expose units prematurely to threats, etc. For a candidate option the coordination requirements and precedence relations are often well-specified. The commander's task is to maximize the coordination in a dynamically evolving and highly stochastic environment where estimates of decision states are uncertain, and where incorrect situation assessments can result in large time delays and/or costly errors.

5. Information-Oriented Decision Tasks

These tasks involve making decisions relative to need, use and dissemination of information. For example, we consider as a "decision" whether to obtain better information or to take action using present information.[4] The trade-offs as always are between urgency and accuracy. Tasks of advising and/or responding can be viewed as information processing within a distributed/decentralized C^2 process. However, at the single operator (node) level, tasks of advising/responding are *decisions*, inasmuch as actions are demanded by the external environment, and local criteria such as maintaining silence may have high utility. On the other hand, a decision whether to transmit or broadcast misleading information may accompany a given action alternative. In general terms, these information-oriented decision tasks are *support* decisions for the primary decision modes of sequencing, resource allocation and scheduling. They are the decisions to get information and to send information, and provide triggering actions for subsequent information processing subtasks.

Finally, the commander should be reviewing the ways in which an enemy may be spoofing him or denying him information, and he should be taking action to develop alternate sources or to counter the enemy actions.

E. Action Implementation Tasks

The tasks associated with response deal with the steps required to implement a chosen course of action. These tasks can conveniently be categorized according to (1) planning, (2) organizing and (3) executing the response. Actual implementation will require the transmission of commands to the effectors within the C^3 system. Planning involves the definition of tasks, resources and schedules required to implement the decision, often including alternatives and contingency effects. Organizing involves the commitment and coordination of resources across dimensions of time, space, and authority hierarchy. Executing is the actual coordinated sequencing of activities associated with the plan. Treating decisions in concert with their plans for implementation enables one to consider human misjudgement. Thus, a decisionmaker may correctly decide to take action but err in either or both of two dimensions: *how* (i.e., the specifics of taking the action), or *when* (i.e., the timing).

F.   Summary

The foregoing subsections have described the tasks and subtasks involved in command and control decisionmaking, as well as the associated opportunities for decision error. Important implications for aiding the decisionmaking process arise directly from these considerations and have been summarized by Wohl (1981b).

VI.   TEAMS OF INTERACTING DECISIONMAKERS

In the foregoing discussion we have concentrated on the single decisionmaker, decision agent, or decisionmaking entity, more or less as though decisions were made by such entities in isolation from one another. But this is almost never the case, especially in large, complex military organizations having multiple functions, missions and broadly capable resources. One decisionmaker must inevitably interact with another, and even when they are well-trained as a team their objectives, situation assessments, and resource needs can often come into conflict. While much has been written in the popular corporate management literature about management (hence, decisionmaking) teams, there is little theory or data to help us to understand and predict for example, the "transient response" of a team's performance when an experienced team member is replaced.

We shall review three recent command and control oriented studies which appear to be moving in the direction of a testable, quantitative theory of team decisionmaking.

1.   Extensions of Organization Theory

Teams of two interacting military decisionmakers (superior and subordinate), Levis (1980) describes the interactions in the following terms:

- The two decisionmakers receive different partitions of the organization's input information; e.g., the subordinate receives detailed

information about a small portion of the environment on which he has to act and the superior receives less detailed information about the "bigger picture."
- Each decisionmaker processes his input in accordance with his own algorithms in order to obtain a partial assessment of the external situation.
- The partial assessments are communicated to each other.
- The superior decisionmaker modifies his situation assessment based on his subordinate's situation assessment input and selects a response which is then communicated to the subordinate decisionmaker as a command.
- The subordinate receives the command input and, on the basis of that plus his modified situation assessment selects a response.
- The subordinate decisionmaker's response then becomes the organization's output.

Levis developed a rigorous mathematical representation of this scenario based on Conant's extension (Conant, 1976) of Shannon's mathematical theory of communication (Shannon and Weaver, 1949) to large-scale systems, and constrained by March's concept of bounded rationality (March, 1978). He shows how each decisionmaker's decision workload is influenced by the situation assessment and by the boundary conditions for rationality. He also shows how admissible and optimal strategies can be computed by this means.

2. "Domules" and "Mutual Models"

Taking a somewhat different approach, Lauer et al. (1981) posed the problem of interacting decisionmakers as follows. In their words, "a domular model of a distributed decisionmaking system consists of a collection of agents (domules), each with its own model and objectives, which coordinate their activities by communicating with one another about their model variables." The decisions of each decisionmaker are constrained to minimize his/its local objective function. Within the bounds of this structure, the authors then examine the effects of three alternative coordination strategies:

- coordination without communication
- coordination with prior communication
- coordination with real-time communication

Again, a rigorous mathematical formulation of this limited team problem was carried out. From a conceptual standpoint, the important addition of this approach is the notion of the composite model for each decision-maker, consisting of his detailed local submodel plus the aggregated sub-models representing each of the other subsystems with which he interacts.

3. The "Expert Team of Experts"

Building on these concepts, Athans (1982) has recently proposed a new "expert team of experts" or ETOE methodology. In this view, each key decisionmaker is considered an expert in his own area of responsibility (e.g., as in the case of the Naval Antisubmarine Warfare Commander); and his expertise is defined by a Principal Expert Model (PEM). Athans notes that a well-functioning organization is more than a team of individual experts. He bases this on the following argument: As a consequence of joint planning, joint training, and availability to all team members of tact-ical, situational, and/or common environmental information, each com-mander or decisionmaker develops a mental picture which represents an aggregated version of the decisionmaking process of his fellow team mem-bers with whom he must coordinate, exploit warfare assets jointly towards a common objective, and sometimes compete for scarce warfare re-sources in pursuit of individual objectives. Athans abstracts this by de-fining for decisionmakers a set of Mutual Expert Models (MEMs), where each MEM is an aggregated version of other decisionmakers' PEMs and all of the MEMs are common to all decisionmakers. The MEMs are de-veloped as a result of joint team training; and it is only in this manner that a joint "team of experts" is transformed into an "expert team of experts" (ETOE).

Note that the totality of the MEMs resident in any military commander or decisionmaker represents an aggregated world view or internal model of the environment and system which through team training and expe-rience is shared among all decisionmakers. This is an important gener-alization of the concept of separate but overlapping detailed individual models posited by Lauer et al. While no quantitative formulation is yet available, Athans has pointed the way for important future research in military decisionmaking.

VII. TASK ISOLATION PROBLEMS AND APPROACHES

A.   Representation/Decomposition Methods

While numerous methods and variants exist for representing and decom-posing a C^3 system and its functions, they can generally be regarded as falling into one of the six categories identified in Table 1.

We have already noted the limitations of any single representational perspective in Section II, as well as the desirability of a unified approach.

Table 1. Categories of Representational/Decompositional Techniques

Type of Relationship Represented	Question Answered	Examples
Physical Constituency	". . . is composed of . . ."	System block diagrams, specification trees
Functional Constituency	". . . involves/requires/ provides . . ."	IDEF$_0$ decompositions, functional block diagrams, function decomposition trees, Interpretive Structural Modeling
Process Constituency	". . . performs the process of . . ."	Mathematical functions, logical operations, rules, procedures
Sequential Dependency	". . . occurs conditional upon . . ."	PERT charts, Petri-nets, scripts, operational sequence diagrams
Temporal Dependency	". . . occurs when . . ."	Event sequences, time lines, scripts, operational sequence diagrams
All of the above	All of the above	Computer simulations, real-world systems

We shall next review briefly key examples in selected categories from Table 1.

Both physical and functional constituency representations generally make use of block diagrams which indicate, name, or otherwise identify for each block, either explicitly or implicitly (i.e., via connectivity), the following:

- functions or processes executed,
- inputs required,
- outputs provided,
- controls applied, and
- resources required.

In addition, each individual block in a diagram can be further decomposed into *its* constituent block diagram, and so on until the level of primitives is reached (e.g., for functional representations, "operator pushes button" or "relay closes" or "message is transmitted," etc.). To handle the added complexity associated with these different levels of description, decomposition trees are often introduced to represent the relationship between levels.

Operational constituency representations generally describe either mathematical or logical operations upon inputs to produce outputs. They can also take the form of explicit rules ("if-then-else-") and procedures.

Temporal and sequential dependency representations nearly always describe functions, processes, activities, and events on a timeline basis in order to indicate either their required sequence or timing or both.

It is of interest to note that sequence or timeline representations of system activities cannot describe feedback loops and their effects; nor can functional or physical block diagrams describe temporal or sequential dependencies. The latter dependencies are extremely important in analyzing or designing C^3 systems in which humans play a central role, especially in determining and allocating human workload in systems. Since time available versus time required for a given process or activity is an important determinant of human workload and human error, time domain descriptions (or their equivalent transforms) of human activities will continue to be indispensible additions to functional, physical, and operational domain representations until such time as a unifying representational technique is developed. Meanwhile, the SHOR paradigm as described in Section V and its mapping onto data processing, information processing and decisionmaking tasks provides a general methodology for the decomposition of a C^2 function into component tasks.

The SHOR paradigm is a major step in going from a qualitative to a quantitative description of a C^2 function provided the tasks so identified are amenable to analysis. However, in adhering to a top-down decomposition, one cannot expect that tasks will appear neatly as isolated entities, awaiting the analyst. Indeed, the current architectural trend in C^3 system design is towards survivability through redundancy and dispersal, while achieving stability through a multilevel command hierarchy.

B.   Organizational Factors

The requirement for a hierarchy of military command stems from a fundamental limit on span of supervision. A manager can successfully take responsibility for not more than five to nine separate activities. Most organization charts, both military and corporate, reflect this limit. As organizations grow in size, they therefore grow in depth of hierarchy. Thus, it is important to remember that the C^2 functions take place at all levels of a command hierarchy, that operator tasks are performed at all levels, and that tasks at any one level involve coordination with those immediately above and below. Consequently, function decomposition is likely to result in a distributed, decentralized hierarchy of operator tasks as shown in Figure 16 where tasks are linked in dependency relations via their outputs, inputs and local objectives. A function may have intra- and inter-level dependencies.

At first sight, such a task network appears to be a formidable web from which to extract a task earmarked for subsequent independent analysis

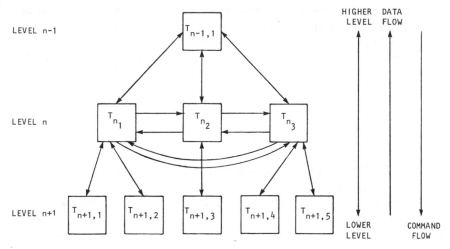

Figure 16. Task Organizational Hierarchy within a Function.

or simulation. Fortunately, there are some fundamental properties of an organizational heirarchy in terms of time constraints, information needs, objectives, and resource requirements that bear on the problem.

- *Time Constraints.* The higher level tasks generally take more time to execute and have a longer time window than the lower level tasks. This is because of their stronger dependence on materiel, maintenance, and personnel resource scheduling/movement. Thus, holding times and processing delays increase upwards in the hierarchy.
- *Information Needs.* The lower level tasks require more informational detail dealing with a smaller number of alternatives/hypotheses, while higher level tasks require less detail and more aggregation. Thus, an aircraft pilot on an interdiction mission will require precise knowledge of threat node locations and characteristics on or near the planned routes, etc. However, the ASOC (air support operations center) responsible for tasking execution and control would only require enough information about enemy threats to allocate aircraft types, defense suppression forces, etc.
- *Objectives.* Commands and information requests flow down the hierarchy, while reporting, acknowledgements, data collection and updating generally flow from lower to higher levels. As a result, inputs from higher level tasks impact set points, resource pools, goals and objectives, and constraints. On the other hand lower level tasks assume the character of clarification, acceptance, implementation and execution.

- *Resource Requirements*. Performance of a C^2 task at any level of the hierarchy requires that a set of resources necessary and sufficient to perform that task be available at the appropriate time and in the appropriate sequence. Tasks at the same or adjacent levels requiring the same or overlapping sets of resources thus may present a potential queuing problem in distributed architectures. Tasks at higher levels tend to require more general-purpose and nonreal-time resources. Tasks at lower levels tend to require more special-purpose and near-real-time resources.

The implication of the above properties in terms of task decomposition, is that *when viewed at level n*, tasks above (level < n) and below (level > n) can be severed from tasks at level n across dimensions of time scale, informational detail, objective, and resource requirement. This is shown in Figure 17. The segregation of interlevel tasks is effected by

- considering lower level tasks as callable in-place procedures that are represented in a simple form. Thus, a lower level information processing task can be treated as returning the requested information to level n, but with a delay and/or reporting inaccuracy. A lower level decision/action can be viewed as having been done under directive, returning only a probability of completion and/or a completion time.

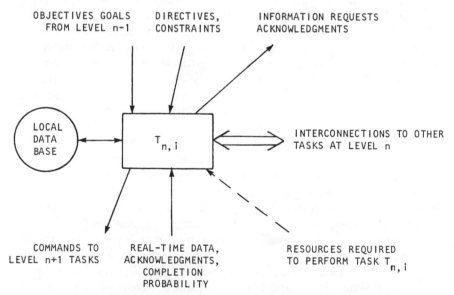

Figure 17. Interlevel Task Decomposition.

- considering higher level tasks as providing the constraints (time, resource, accuracy) and objectives needed for the execution of the level n task. Thus, the higher level provides the executive commands, that translate into the utilities, option alternatives, local cost functional criteria, and constraints for the task construct.

It should be noted that dynamic interlevel coordination is still presumed in the above segregations. For example, higher level constraints may change slowly with time, while lower level data inputs, delays, and performance characteristics may change dynamically. What we do *not* have to consider at level n are the *mechanisms* of the dynamics *internal* to those tasks at other levels.

C. Task Isolation

The above discussion is germane to the problem of interlevel dependencies and hierarchical representation/decomposition. The second problem, that of extracting a task from its intralevel dependencies, is more formidable. Here, tasks are coupled in time frame dynamics, information needs and aggregation, etc. The tasks may be of different type, e.g., data processing, information processing or decisionmaking, and are often executed in parallel. (Tasks performed in time sequence exhibit far less difficulty in their decompositions than do task linked in a dynamic information/decision feedback structure. Time sequence decomposition techniques are standard practice in human factors analysis.) In the cases where tasks exhibit high intralevel coupling we can

1. describe the composite team task as a single task if suitable technology can be brought to bear on the situation, or
2. describe analytically the individual tasks and integrate these parts using decentralized or distributed theories.

As an example, the second approach appears to be feasible when analyzing a distributed multinode AAA system. Each node is at the same hierarchy level and commanders may or may not share information. Node directives emanate from a higher level C^2 node, while local commanders provide supervisory control over their own lower level tasks (tracking, implementation of firing doctrine, etc.). A similar hierarchy may exist within a SAM system. Both threat networks are themselves coordinated/managed at a higher command level into a threat. Different cluster types (e.g., long-range versus short-range air defense) merge at even higher levels, and so on.

Another reason for choosing the second approach which relies on task description followed by merging (rather than the reverse process), is that the loss or nonperformance of a task can be handled with less difficulty. It is not necessary to redo the entire level description, but only to modify the merging logic for the remaining tasks.

As a final comment, we tend to view a task as being defined to a level of detail where it is performed by a single human. In many situations, however, a single task may be performed jointly by a team of N humans or by a human and a computer. One approach is to view such tasks as being performed by a single (pseudo) operator with N-fold capacity. When a task is relegated to a computer, interest focuses not on the individual task, but on the question of how human and computer tasks should interact to form a more perfect union.

VIII.   CONCLUSIONS AND IMPLICATIONS

The objectives of this chapter were to summarize the current state of knowledge regarding human decision processes in military command and control and to identify critical research areas. It is now possible to identify and decompose the key decision activities in a command and control process in terms of the Stimulus-Hypothesis-Options-Response (SHOR) paradigm. Using this structure, it is evident that much effort has been devoted to the problem of option evaluation in the decisionmaking literature, but there is a paucity of research dealing either with option generation (i.e., planning) or hypothesis formation and evaluation (i.e., situation assessment). With few exceptions, the relevance of existing psychological and organizational research is minimal. Some progress has been made in the mathematical representation of a commander's mental model and of the interaction between a commander and his subordinate. However, we must conclude that the capability for quantitative formulation, representation and analysis of both individual and team military decision processes is still in its infancy. While the need for a theory of command and control has been well documented, it must await further results in the theory of military decisionmaking.

NOTES

1.   A correct assessment that a certain object is a hostile reentry vehicle is of minimal value after it has impacted and detonated its warhead!

2.   Depending on the amount of detail included, TAC ASSESSOR may run significantly slower than real time.

3.   Here, and in what follows, the response tasks are deemphasized.

4.   We can always define an action alternative as "obtain more information." However, the decision to seek more information is generally made when evaluating options, not when evaluating ENG.

REFERENCES

Alberts, D. S. C^{2}I assessment. In MITRE, 1980.

Aldrich, J. R. Application of knowledge-based approaches to C^{2}I processes. In Athans, et al., 1980.

Athans, M. (Ed.), *Proceedings of the First MIT/ONR Workshop on Distributed Information and Decision Systems Motivated by C^3 Problems.* Laboratory for Information and Decision Systems. Cambridge, MA: MIT, 1978.

Athans, M. (Ed.), *Proceedings of the Second MIT/ONR Workshop on Distributed Information and Decision Systems Motivated by C^3 Problems.* Laboratory for Information and Decision Systems. Cambridge, MA: MIT, 1979.

Athans, M. System theoretic challenges and research opportunities in military C^3 systems. *Proceedings of the 19th IEEE Conference on Decision and Control,* Albuquerque, New Mexico, 1980.

Athans, M. The expert team of experts approach to command and control organizations. *IEEE Control Systems Magazine,* (September), 1982.

Athans, M., Davenport, W. B. Jr., Ducot, E. R., and Tenney, R. R. (Eds.), *Proceedings of the Fourth MIT/ONR Workshop on Distributed Information and Decision Systems Motivated by C^3 Problems.* Laboratory for Information and Decision Systems. Cambridge, MA: MIT, 1981.

Athans, M., Ducot, E. R., and Tenney, R. R. (Eds.), *Proceedings of the Third MIT/ONR Workshop on Distributed Information and Decision Systems Motivated by C^3 Problems.* Laboratory for Information and Decision Systems. Cambridge, MA: MIT, 1980.

Brandenberg, R. L. Naval warfare environment simulation. In MITRE, 1980.

Bruner, J. S., Goodnow, J. J., and Austin, G. A. *A Study of Thinking.* New York: Wiley, 1956.

Conant, R. C. Laws of information which govern systems. *IEEE Transactions on Systems, Man, and Cybernetics SMC-6*(4), 1976.

Engelman, C., Bergrand, S. H., and Bischoff, M. KNOBS: An experimental knowledge-based tactical air mission planning system and a rule based aircraft identification simulation facility. *Proceedings of the Sixth International Joint Conference on Artificial Intelligence,* Tokyo, Japan, Volume I, 1979, pp. 247–249.

Everett, R. R. Lanchester and C^3. Described in Wohl, et al., 1981.

Freedy, A., et al. Adaptive decision aiding in tactical operations: An application to airborne ASW. *Naval Research Reviews XXXII* (2), 1980.

Johnston, S. C. An interactive computer aiding system for group decisionmaking. In Athans, et al., 1980.

Kahneman, D., Slovic, P., and Deese, J. *General Psychology.* Boston, MA: Allyn and Bacon, 1975.

Kahneman, D. and Tversky, A. Subjective probability: A judgement of representativeness. *Cognitive Psychology 3:*430–454, 1972.

Kelley, C. W. Decision analysis and decision aids. In MITRE, 1980.

Klein, G. A. *Training Requirements Analysis for Tactical Domains.* Yellow Springs, Ohio: Klein Associates, Inc., 1981.

Lauer, G. S., Sandell, N. R. Jr., and Tenney, R. R. Domular modeling of C^2 system: Preliminary assessment. Burlington, MA: ALPHATECH, Inc., Report TR-114, 1981.

Lawson, J. A unified theory of command control. In *Proceedings, 41st Military Operations Research Symposium,* Fort McNair, Washington, D.C., 1978.

Lawson, J. The state variables of a command and control system. In MITRE, 1980a.

Lawson, J. Command and control as a process. In *Proceedings, 19th IEEE Conference on Decision and Control,* Albuquerque, New Mexico, 1980b.

Leedom, D. K. The simulation of tactical decision processes. In Athans, et al., 1980.

Levis, A. H. On organizational forms and C^2 system structure. Cambridge, Massachusetts, Laboratory for Information and Decision Systems, MIT, 1980.

March, J. G. Bounded rationality, ambiguity, and the engineering of choice. *Bell Journal of Economics* (March), 1978.

Meyer, M. E. (Ed.), *Foundations of Contemporary Psychology*. New York: Oxford Press, 1979.

Miller, H. G. Sea-based air C^3 effectiveness. In MITRE, 1980.

Miller, H. G. C^3 impact on force coordination. In Athans, et al., 1980.

Miller, L. W., Kaplan, R. J., and Edwards, W. JUDGE: A value-judgement-based tactical command system. Santa Monica, California, RAND Corporation, Report No. RM-5147-PR, 1967.

Miller, L. W., Kaplan, R. J., and Edwards, W. JUDGE: A laboratory evaluation. Santa Monica, California, RAND Corporation, Report No. RM-5547-PR, 1968.

MITRE. *Proceedings for Quantitative Assessment of Utility of Command and Control Systems*. Office of the Secretary of Defense with the Cooperation of the MITRE Corporation, C^3 Division, McLean, Virginia, The MITRE Corporation, Report MTR-80W00025, 1980.

Modrick, J. A. Decision support in a battlefield environment. In Thrall, et al., 1976.

Pattipati, K. R., and Kleinman, D. L. A survey of the theories of individual choice behavior. Storrs, Connecticut, University of Connecticut Technical Report EECS TR-79-12, 1979.

Rona, T. P. Generalized countermeasure concepts in C^3. In Athans, 1979.

Rona, T. P. Survey of C^3 assessment concepts and issues. In MITRE, 1980.

Sage, A. P. Behavioral and organizational considerations in the design of information systems and processes of planning and decision support. *IEEE Transactions on Systems, Man, and Cybernetics SMC-11*(9):640–678, 1981.

Sandell, N. R., Jr Distributed decisionmaking processes in C^3. In Athans, et al., 1980.

Shannon, C. E. and Weaver, W. *The Mathematical Theory of Communication*. Urbana: University of Illinois, 1949.

Tenney, R. R. Modeling the C^3 decision process. In Athans, et al., 1980.

Tenney, R. R. and Sandell, N. R. Jr. Structures for distributed decisionmaking. *IEEE Transactions on Systems, Man and Cybernetics SMC-11*:517–526. 1981a.

Tenney, R. R. and Sandell, N. R. Jr. Strategies for distributed decisionmaking. *IEEE Transactions on Systems, Man, and Cybernetics SMC-11*:527–535, 1981b.

Thrall, R. M., Tsokos, C. P., and Turner, J. C. *Proceedings of the Workshop on Decision Information for Tactical Command and Control*. Houston, TX: R. M. Thrall and Associates, 1976.

Tversky, A. and Kahneman, D. Judgement under uncertainty: Heuristics and biases. *Science 185*:1124–1131, 1974.

Van Trees, H. L. Keynote address—Conference workshop on quantitative assessment of the utility of command and control systems. In MITRE, 1980.

Wason, P. The psychology of deceptive problems. *New Scientist 63*:382–385, 1974.

Welch, J. A., Jr. State of the art of C^2 assessment. In MITRE, 1980.

Welch, J. A., Some random thoughts on C^3. In Athans et al., 1981.

Wohl, J. G. Battle management decisions in Air Force Tactical Command and Control. Bedford, Massachusetts, MITRE, Report M79-233, 1979.

Wohl, J. G., (Ed.), Information processing and decisionmaking for battle managers. In *Proceedings of the Third Annual Symposium on C^2*. Bedford, MA: MITRE, 1980a.

Wohl, J. G. (Ed.), *Modeling and Simulation of Avionics Systems and Command, Control and Communications Systems*. AGARD Conference Proceedings, Number 268, 1980b.

Wohl, J. G. Rate of change of uncertainty as an indication of command and control effectiveness. *Proceedings, 47th Military Operations Research Society Symposium*, Washington, D.C., 1981a.

Wohl, J. G. Force management requirements for Air Force Tactical Command and Control. *IEEE Transactions on Systems, Man, and Cybernetics SMC-11*:618–639, 1981b.

Wohl, J. G., Gootkind, D., and D'Angelo, H. *Measures of Merit for Command and Control*. Bedford, MA: MITRE, Report No. MTR-8217, 1981.

Zracket, C. A. Issues of C^2 R&D evaluation. In MITRE, 1980.

AUTHOR INDEX

309

SUBJECT INDEX

Research Annuals in
Computer Science

Advances in Automation and Robotics
  Edited by George N. Saridis, *Dept. of Electrical, Computer and Systems Engineering, Rensselaer Polytechnic Institute*

Advances in Computer-Aided Engineering Design
  Edited by Alberto Sangiovanni-Vincentelli, *Dept. of Electrical Engineering and Computer Science, University of California, Berkeley*

Advances in Computer Methodology for Management
  Edited by Roman V. Tuason, *The Concord Group Inc.*

Advances in Computer Software Applications
  Edited by C. Keith Conners, *Dept. of Psychiatry, Children's Hospital, National Medical Center*

Advances in Computer Vision and Image Processing
  Edited by Thomas S. Huang, *Coordinated Science Laboratory, University of Illinois*

Advances in Computing Research
  Edited by Franco P. Preparata, *Coordinated Science Laboratory, University of Illinois*

Advances in Flexible Manufacturing Cells
  Edited by Paul K. Wright, *Dept. of Mechanical Engineering, Carnegie Mellon University*

Advances in Geophysical Data Processing
  Edited by Marwan Simaan, *Interactive Computing Laboratory, Dept. of Electrical Engineering, University of Pittsburgh*

Advances in Large Scale Systems: The Theory and Applications
  Edited by Jose B. Cruz, Jr., *Coordinated Science Laboratory, University of Illinois*

Advances in Man-Machine Systems Research
  Edited by William B. Rouse, *Center for Man-Machine Systems Research, Georgia Institute of Technology*

Advances in Statisical Analysis and Statistical Computing: Theory and Application
  Edited by Robert S. Mariano, *Dept. of Economics, University of Pennsylvania*

Advances in Software Engineering
  Edited by Stephen S. Yau, *Dept. of Electrical Engineering and Computer Science, Northwestern University*

Please inquire for detailed brochure on each series.

 **JAI PRESS INC.**